HIGH-SPEED CLOCK NETWORK DESIGN

High-Speed Clock Network Design

by

Qing K. Zhu

Intel Corporation, T-RAM Inc., U.S.A.

KLUWER ACADEMIC PUBLISHERS

BOSTON / DORDRECHT / LONDON

A C.I.P. Catalogue record for this book is available from the Library of Congress.

ISBN 1-4020-7346-1

Published by Kluwer Academic Publishers,
P.O. Box 17, 3300 AA Dordrecht, The Netherlands.

Sold and distributed in North, Central and South America
by Kluwer Academic Publishers,
101 Philip Drive, Norwell, MA 02061, U.S.A.

In all other countries, sold and distributed
by Kluwer Academic Publishers,
P.O. Box 322, 3300 AH Dordrecht, The Netherlands.

Printed on acid-free paper

Printed in the Netherlands.

Table of Contents

Preface

The author would like to thank Mr. Mark D. Jongh at Kluwer Academic Publishers for giving the opportunity to publish this book. He also appreciates the help from Peter Chau and Huiling Song for their reviewing and encouragements during the book writing.

This book is a collection of design concepts, techniques and research works from the author for clock distribution in microprocessors and high-performance chips. It is organized in 11 chapters as follows. Chapter 1 provides an overview to the design of clock networks. Chapter 2 specifies the timing requirements in digital design. Chapter 3 shows the circuits of sequential elements including latches and flip-flops. Chapter 4 describes the domino circuits, which need special clock signals. Chapter 5 discusses the phase-locked loop (PLL) and delay-locked loop (DLL), which provide the clock generation and de-skewing for the on-chip clock distribution. Chapter 6 summarizes the clock distribution techniques published in the state-of-the-art microprocessor chips. Chapter 7 describes the CAD flow on the clock network simulation. Chapter 8 gives the research work on low-voltage swing clock distribution. Chapter 9 explores the possibility of placing the global clock tree on the package layers. Chapter 10 shows the algorithms of balanced clock routing and wire sizing for the skew minimization. Chapter 11 shows a commercial CAD tool that deals with clock tree synthesis in the ASIC design flow. The glossary is attached at the end of this book. The clock network design is still a challenging task in most high-speed VLSI chips, since the clock frequency and power consumption requirements are increasingly difficult to meet for multiple clock networks on the chip. Many research works and industry examples will be shown in this area to continually improve the clock distribution networks for future high-performance chips.

Chapter 1 Introduction

Clock distribution is one of the limiting factors for the high frequency chip design. Device technology improvement, such as deep-submicron with faster transistors, can only marginally solve the clock distribution problem because the interconnect delay becomes the significant factor in the clock cycle time. Section 1.1 provides the introduction to clock frequency and power issues. Section 1.2 explains the sources of the clock skew and clock jitter, which are two major design issues in the clock network design. Section 1.3 shows the PVT effects on the clock skew. Section 1.4 describes the clock buffer design. Section 1.5 discusses the power supply and reliability for clock distribution. Section 1.6 demonstrates the design complexity of the clock distribution network by using a microprocessor example. Section 1.7 provides the summary to this chapter.

1.1 Clock Frequency and Power Consumption

Figure 1-1(a) and Figure 1-1(b) show the trends of interconnect delay versus gate intrinsic delay and microprocessor clock frequency. A clock frequency has been achieved at 2.5GHz in Intel Pentium-IV chips. Chip technology improvement such as smaller feature size, larger chip area, and increased component density are increasing the difficulties of the clock distribution, since they usually result in higher series interconnect resistance and higher clock loads.

Insertion of multiple intermediate levels of clock buffers helps in the transition time and power reduction for the clock tree. But it cannot resolve the clock skew problem, because the clock skew is caused by the process variations in the clock buffers. Extensive research has been performed on the RC delay balance by equating the delay along the branches of the clock tree. H-tree design is widely used in the industry for achieving the equal path lengths based on the recursive H tree structure [2,3]. IBM's Power-PC and DEC's Alpha microprocessor chips both use the H-tree design for global clock distribution [4,5]. A limitation of the H-tree is that the clock sinks should be placed in symmetric or uniform locations. Clock grids or clock trunks are also used in the industry with the penalty of taking a large wiring area and causing much more load capacitances [6]. Balanced clock tree construction algorithms for the arbitrary locations of clock sinks (to eliminate the H-tree limitation) can be found in academic research works by Jackson and Kuh (Method of Means and Medians) [7], Kahng, Cong, and Robins (Recursive Geometric Matching Method) [8], Tsay (Elmore Delay Matching Method) [9], Chao, Hsu, Ho, Edahiro, Boese, and Kahng (Deferred-Merge Embedding Methods) [10,11,12], Chou and Cheng (Simulated Annealing Method) [13], Zhu, Dai, Kahng, and Tsao (Planar Equal Path Length Clock Tree Methods) [14,15,16].

Not only is skew an important consideration when designing a clock network, but other factors must be taken into account as well. For instance, clock rise and fall time is proportional to both line resistance and load capacitance, so it is desirable to reduce both of these characteristics. References [24-25] discussed the statistical models for IC chip circuit simulation. Reference [26] considered the delay matching of two clock buffers in pull-up and pull-down edges with process variations. References [27-29] considered the process variations in clock buffer sizing and wire sizing. Reference [30] showed the general principles behind the clock buffer delay and size optimization. Reference [31] showed an analytic way to determine the clock buffer size.

In order to reduce the power consumption, it is useful to adopt the clock gating whenever possible in the clock distribution network. In addition, reducing gate and line capacitances of the clock network will help the power reduction. The clock network itself may take up to 50% of the chip total power consumption, as illustrated in Figure 1-2, based on the data from Intel.

1.2 Sources of Clock Skew and Clock Jitter

With increasing clock frequency, clock inaccuracy has become a large percentage of the clock period. It is critical that one accounts for the clock inaccuracy in the minimum and maximum path delay calculations. The clock inaccuracy is basically a spatial and temporal phenomenon, causing one clock edge not to arrive at a sampling point at its nominal time, measured relative to another clock edge. The clock inaccuracy consists of two components: the clock skew and the clock jitter. Clock skew represents the spatial separation effect of the clock inaccuracy, whereas clock jitter represents the temporal separation effects of the clock inaccuracy. Figure 1-3 illustrates the clock skew that is the path delay difference of clock signals from the common clock root (the output of the on-chip phase-locked loop) to two clocked elements. The clocked elements can be the latch, flip-flop, domino circuit, etc. There are systematic and random effects that contribute to the clock skew as follows [58]:

- Variations in effective channel lengths of devices across the die. This is caused by global in-die variations resulting from lens distortions, wafer planarity, stepper accuracy etc. Also local effects like device proximity and some other random phenomena (e.g. dopant infiltration of the channel) result in channel length variations.

- Threshold voltage Vt variation across the dieWider devices has smaller variations in Vt compared to narrower devices. Typical variation could be as high as 50mV for narrow devices.

- Inter-layer dielectric (ILD) thickness variation. This causes the interconnect capacitance to vary, resulting in delay variation in interconnect dominated paths.

- Design errors. These arise from human errors like mistakes in modelling the circuit, choice of simulation step size, etc., and from the tool related errors in RC parasitic extraction, simulator accuracy, etc.

- Supply voltage variation across the die. This depends on power grid design, proximity to high activity and large devices, etc.

- Temperature difference across the die. It is very likely to have 5-10oC across a microprocessor chip.

- Signal coupling to neighbouring lines. One path may be highly coupled to neighbouring lines compared to another, causing the clock skew between two paths.

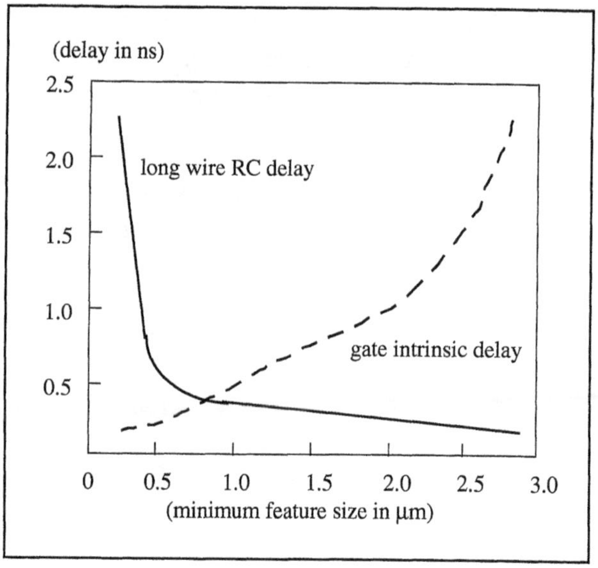

(a) Source: TI Corporation, Shin-Puu Jeng

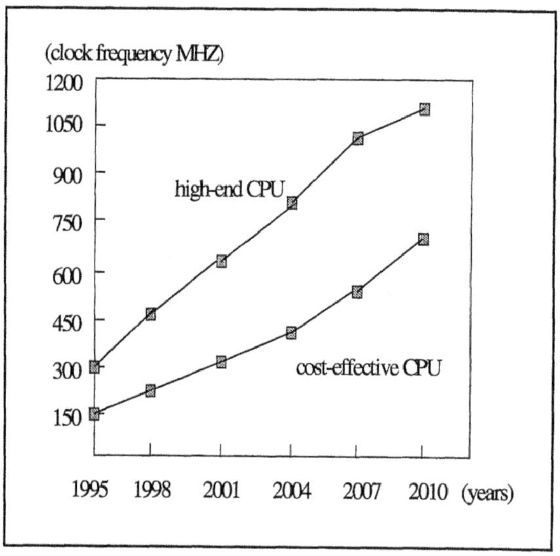

(b) Source: SIA Road Map

Figure 1-1. Interconnect Delay and CPU Clock Frequency Trends.

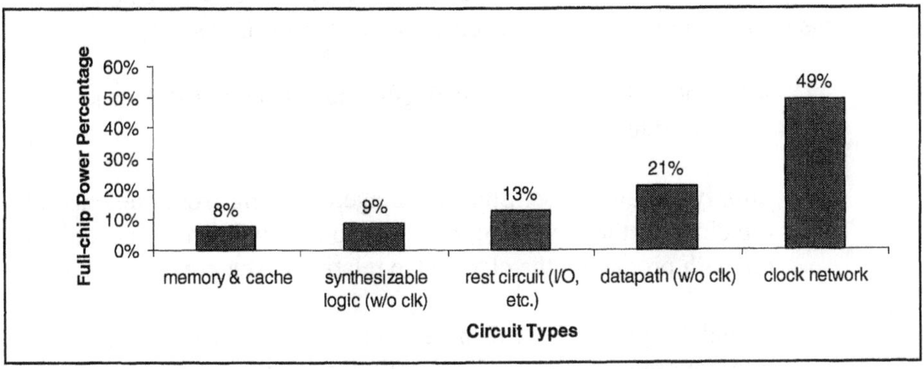

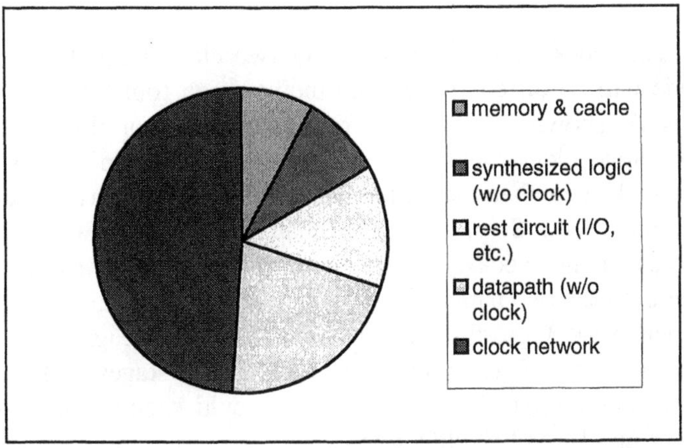

Figure 1-2. Power Consumption in Pentium-II® Microprocessor.

Clock jitter, on the other hand, represents the inaccuracy in two clock edges that occur at different times. The timing diagram shown in Figure 1-4 illustrates this phenomenon. Assume that frequency of the clock is 500 MHz – so ideally, each rising edge of ClkA is separated from the next rising edge by 2ns. However, in reality, due to factors like power supply noise, signal coupling, etc. the next rising edge may appear earlier or later than 2ns. Note that if the jitter ClkA signal is monitored on an oscilloscope triggered on an ideal 500MHz clock, the clock edges will appear as thick bands. This thickness of the bands represents the excursion of the clock edges from the nominal timing and is referred to as the peak-to-peak jitter. Although the example in Figure 1-4 shows the clock jitter on the same clock line (i.e. ClkA), the concept is equally applicable to two different clocks – for example ClkA and ClkB. The clock jitter will be on top of the clock skew between two clock signals. Some of the major sources of the clock jitter are as follows [58].

- Inherent PLL jitter. The clock generated from the PLL has some inherent jitter.
- Supply voltage noise. Supply voltage changing from one edge of the clock to the next edge, resulting in a slowdown or speedup of the path producing the clock edge relative to the previous cycle.
- Signal coupling noise. If there is intermittent activity on a signal line coupling to a clock line while the clock is switching, every time the signal line switches, there will be a delay difference between the clock paths.

The clock skew requirement for two clock paths from the same root node will be different based on the common root from two clocked elements. As shown in Figure 1-5 for a microprocessor clock distribution tree, two local clocks with gclk-4 as the first common node will have smaller clock skew than that of two local clocks, which have gclk-8 as the first common node. Since each stage that is different between two paths contributes to the clock skew, reducing the number of stages that are different between two paths will reduce the clock inaccuracy. Tie together clocks which are logically the same when they are geographically close and the common node is no more than two or three stages back. Minimize the signal coupling to clock lines. This would require increasing line spacing or insertion of shielding wires.

1.3 On-Die Variations and Clock Skew Impacts

Distributing the clock with the minimal skew is a major concern in a high-speed digital hardware design. Clock skew is due to the variations in the arrival time of a clock signal at various parts of a digital circuit. Even with the equal load for two clock buffers in the layout, there is still clock skew caused by the on-die variations of the process, temperature and supply voltage (PVT variations) across the VLSI chip. We provide the experimental results for the clock skew impacts due to PVT variations for the on-chip clock buffers.

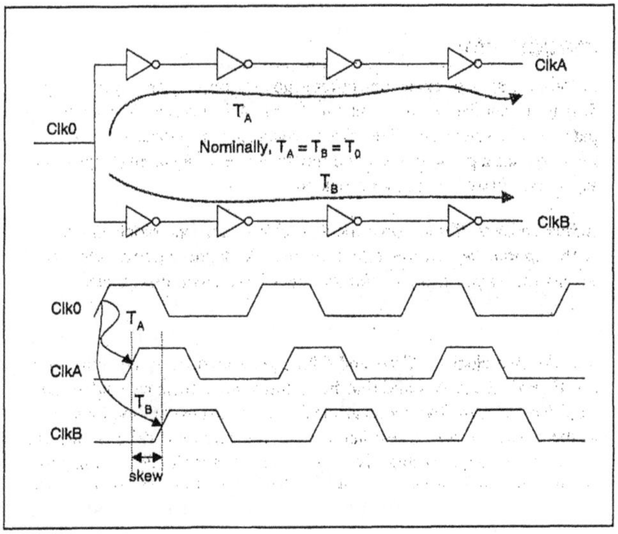

Figure 1-3. Clock Skew Definition [58].

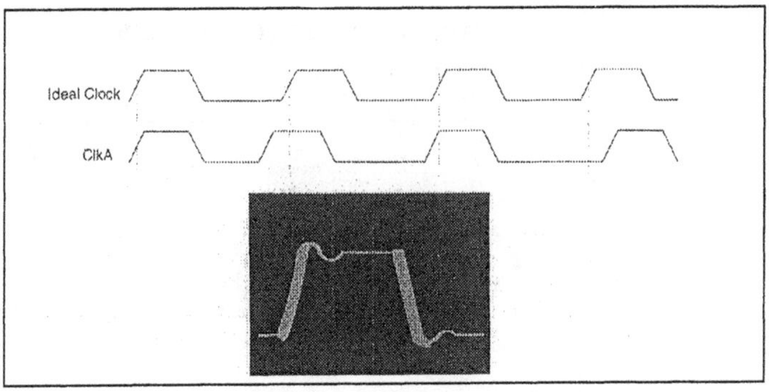

Figure 1-4. Clock Jitter Definition [58].

Figure 1-6 shows a clock buffer structure. It consists of an Inverter and a NAND gate for the clock gating purpose, which accepts an Enable signal together with the input clock. Tables 1-1, 1-2, and 1-3 show the simulation results of the clock buffer delay and output slope in various supply voltages, temperatures, and process corners for a 0.18μm process. The clock buffer delay is defined by the 50%Vdd of the input clock waveform to the 50%Vdd of the output clock waveform through the clock buffer. The output slope is defined as 10%Vdd to 90%Vdd for the output clock waveform of the clock buffer. We can observe that the delays and output slopes are significantly changed in various PVT conditions, which are transferred into the clock skew for two clock buffers in different PVT conditions. In order to simulate the clock skew across the chip, the modelling of the PVT variations is critical and we discuss the statistical method in this section as follows.

To avoid the worst-case corner simulations that may result in unrealistic skew numbers, the statistical models in the simulation are necessary to model the PVT conditions [49]. The parameter variations allow the circuit simulator to predict the actual circuit response to the extremes of the manufacturing process. The possible PVT parameters in a 0.18μm process are listed in Table 1-4. Parameters are chosen based on the available historical data collected either during fabrication or electrical test. For example, the polyCD skew parameter XL is collected during fabrication. This parameter is usually the most important skew parameter for a MOS process. The skew parameters are model parameters. They are used most often for transistor models, but also applicable to passive elements. A polysilicon might be drawn at 2μm dimension and shrunk to 1μm (LMLT/WMLT), with a physical size 0.9μm (XL/XW) and an effective size 0.8μm (LD/WD), as illustrated in Figure 1-7.

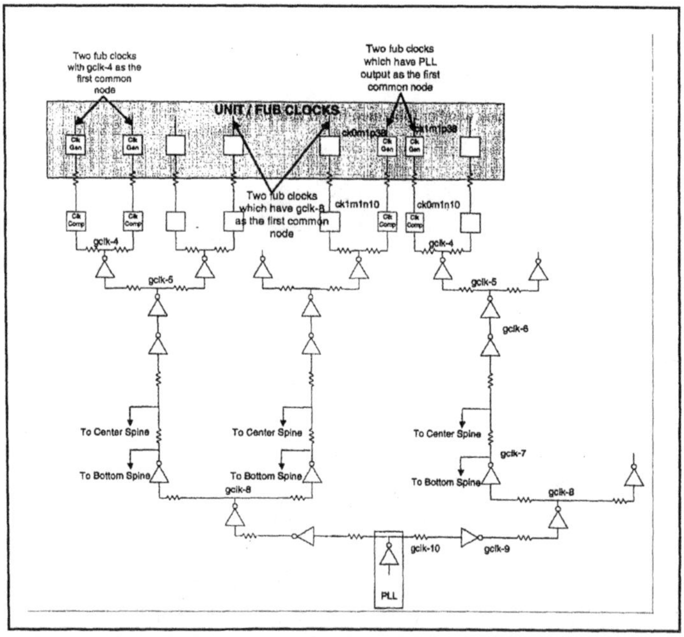

Figure 1-5. Clock Distribution Hierarchy [58]. The PLL is the root of the clock tree.

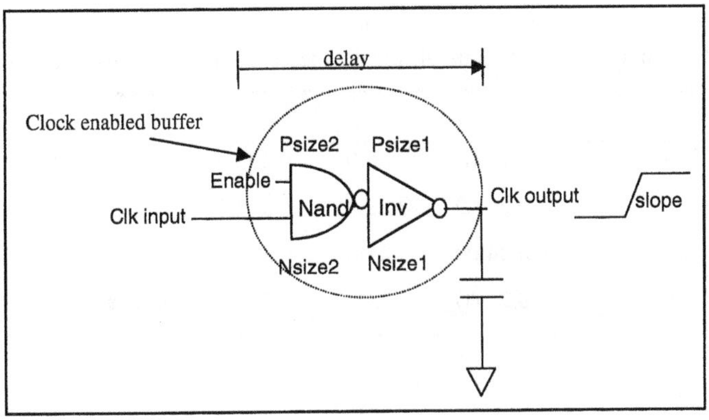

Figure 1-6. Clock Buffer Structure.

Table 1-1. Clock Buffer Delays and Output Slopes in Different Supply Voltages (0.18μm, ss process corner, 110°C).

Voltage (V)	*1.4*	*1.5*	*1.6*	*1.7*	*1.8*	*1.9*
Buffer Delay (ns)	0.45 (r)	0.40 (r)	0.37 (r)	0.34 (r)	0.32 (r)	0.30 (r)
	0.47 (f)	0.42 (f)	0.40 (f)	0.37 (f)	0.35 (f)	0.33 (f)
Output Slope (ns)	0.42 (r)	0.39 (r)	0.36 (r)	0.34 (r)	0.32 (r)	0.31 (r)
	0.35 (f)	0.33 (f)	0.31 (f)	0.30 (f)	0.29 (f)	0.28 (f)

Table 1-2. Clock Buffer Delays and Output Slopes in Different temperatures (0.18μm, ss process corner, 1.8V).

Temperature (°C)	0	25	50	75	100	125
Buffer Delay (ns)	0.30 (r)	0.31 (r)	0.31 (r)	0.32 (r)	0.33 (r)	0.34 (r)
	0.33 (f)	0.34 (f)	0.34 (f)	0.35 (f)	0.35 (f)	0.36 (f)
Output Slope (ns)	0.29 (r)	0.30 (r)	0.31 (r)	0.32 (r)	0.33 (r)	0.34 (r)
	0.25 (f)	0.26 (f)	0.27 (f)	0.29 (f)	0.30 (f)	0.32 (f)

Table 1-3. Simulation Results in Different Process Corners (0.18μm, 1.8V, 110°C).

Process corners	FF	TT	FS	SF	SS
Buffer Delay (ns)	0.22 (r)	0.26 (r)	0.23 (r)	0.29 (r)	0.32 (r)
	0.26 (f)	0.29 (f)	0.32 (f)	0.28 (f)	0.35 (f)
Output Slope (ns)	0.24 (r)	0.27 (r)	0.29 (r)	0.26 (r)	0.32 (r)
	0.21 (f)	0.25 (f)	0.23 (f)	0.27 (f)	0.29 (f)

Table 1- 4. PVT Parameters in Statistical Device Modelling [49].

Parameters	*Variations*
XL	Polysilicon CD (critical dimension of poly layer representing the difference between drawn and actual sizes)
XWn, XWp	Active CD (critical dimension of active layer representing the difference between drawn and actual sizes)
TOX	Gate oxide thickness
RSHp, RSHn	Active layer resistivity (PMOS/NMOS)
DELVTOn, DELVTOp	Threshold voltage variation
TEMP	Temperature variation
VDD, VSS	Vdd and Vss variations

Table 1-5 shows whether deviations are added to or subtracted from the worst simulation corners. The following is the statistical simulation methodology using the HSPICE simulator [52]. The simulator based on Monte Carlo analysis uses a random number generator to create the Gaussian parameter distribution or uniform parameter distribution. To set up a Monte Carlo analysis, use the .PARAM statement in the HSPICE simulation netlist for Gaussian or uniform statistical distribution. In the .MEASURE statement, calculate the output mean, variance, sigma, and standard deviation. A new random variable is calculated each time when a parameter is used. The actual element value will vary by the element distribution (Gaussian or Uniform). A Monte Carlo distribution also can be assigned to a model keyword, and all elements that share the model use the same keyword value. For example, the MOSFET channel length varies from transistor to transistor by a small amount corresponding to the die distribution.

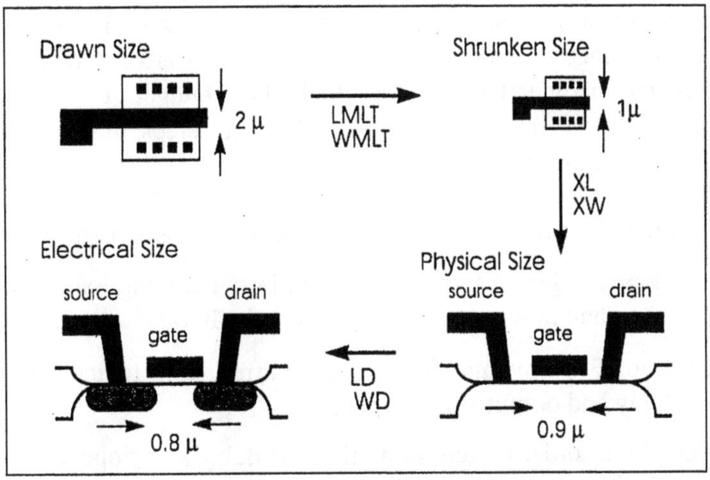

Figure 1-7. Scaling from Drawn Size to Electrical Size [52].

Table 1-5. Statistical Deviations from Worst-Case Process Corners.

Type	*Parameter*	*Slow corner*	*Fast corner*
NMOS	XL	+	-
	RSH	+	-
	DELVTO	+	-
	TOX	+	-
	XW	-	+
PMOS	XL	+	-
	RSH	+	-
	DELVTO	-	+
	TOX	+	-
	XW	-	+

1.4 Clock Buffer Circuit Design

Table 1-6 shows the clock buffers with different transistor sizes for driving various capacitance loads in a 0.18μm microprocessor. All the clock buffers have the delays within an error window, so the clock skew will be bounded by the required error in a specified clock frequency target. Shown in Table 1-6(a), CB represents the clock buffer structure with an inverter followed by another inverter. Shown in Table 1-6(b), SQ represents the structure with a NAND gate and an inverter as shown in Figure 1-6(a). SQ is the structure to provide the clock enable signal for the clock gating purpose. Table 1-6 also shows the minimum (Cmin) and maximum (Cmax) capacitance range for each clock buffer to drive. The capacitance range is important to satisfy the delay error window for each clock buffer in order to meet the clock skew budget. The clock buffer sizing design is done using the following steps for this 0.13μm microprocessor project [49]:

- Step (1): decide the clock buffer delay and output slope targets.
- Step (2): decide the overall clock buffer driving load capacitance range and how many buffer sizes in the library.
- Step (3): simulate the selected buffer sizes to meet the buffer delay and output slope targets.

For Step (1), in order to determine the best delay and slope targets (*D*, *S*) for clock buffer sizes, the sensitivity trends of the clock skew and buffer size will be studied. The results in a 0.13μm process are shown in Figure 1-8 [49]. Within a clock skew upper bound, we select the largest possible delay and slope targets (*D*, *S*) to reduce the buffer size. We simply shift the initial buffer delay (*D*) to higher end in Figure 1-8(b), until the skew is no longer less than the upper bound (S_{max}). In our case, as shown in Figure 1-8(b), the 145.6ps delay target and 68.9ps output slope target are selected to meet the clock skew upper bound of 50ps. The selected target of 145.6ps results in about 20% smaller clock buffer size compared to the result using the initial delay target of 112ps.

Table 1-6(a). CB Type Clock Buffer Library (0.18μm Process).

Buffer Size	C_{min} (pF)	C_{max} (pF)	C_{max} - C_{min} (pF)
S5	0.010910	0.015064	4.7fF
S4	0.015064	0.019680	4.6fF
S3	0.019680	0.025103	5.4fF
S2	0.025103	0.030821	5.7fF
S1	0.030821	0.040073	9.3fF
RS	0.040073	0.053344	13.3fF
R1	0.053344	0.073852	10.5fF
R2	0.073852	0.100036	26.2fF
R3	0.100036	0.142250	42.2fF
R4	0.142250	0.195927	53.7fF
R5	0.195927	0.280304	84.4fF
R6	0.280304	0.365905	85.6fF
R7	0.365905	0.495819	129.9fF
R8	0.661987	1.001709	339.7fF

Table 1-6(b). SQ Type Clock Buffer Library (0.18μm Process).

Buffer Size	C_{min} (pF)	C_{max} (pF)	$C_{max} - C_{min}$ (pF)
S5	0.010050	0.014644	4.6fF
S4	0.014644	0.018159	3.5fF
S3	0.018159	0.023771	5.6fF
S2	0.023771	0.027757	4.0fF
S1	0.027757	0.038646	10.9fF
RS	0.038646	0.049211	10.6fF
R1	0.049221	0.073433	24.2fF
R2	0.073433	0.100708	27.3fF
R3	0.100708	0.138222	37.5fF
R4	0.138222	0.179931	41.7fF
R5	0.179931	0.262512	82.6fF
R6	0.262512	0.346099	83.6fF
R7	0.346199	0.453857	107.6fF
R8	0.346199	0.453857	107.7fF
R9	0.617004	0.956005	339.0fF

For Step (3), an optimization circuit for the buffer sizes is needed. One set of these optimization circuits is shown in Figure 1-9. Figures 1-9(a) and (b) are for the transistor size optimization of CB type buffers, and Figures 1-9(c) and (d) are for SQ type. The circuit can be used for each buffer size according to the load range in order to meet the selected delay and output slope targets in addition to minimizing the transistor sizes. The difference between Figure 1-9(a) and Figure 1-9(b) is that Figure 1-9(a) is suitable for buffer sizes in high loads, while Figure 1-9(b) is suitable for buffer sizes in light loads. In the light load case, it may be very hard to meet the delay and output slope targets; therefore a dummy load is added between the first and second stage inversions in the buffer.

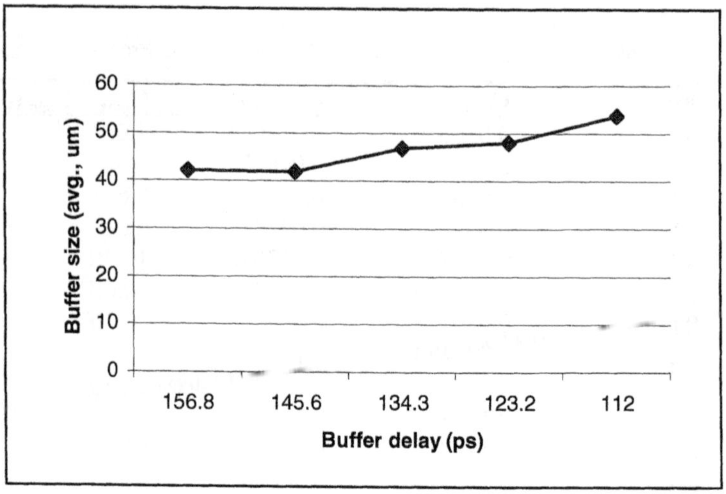

(a) Buffer size vs. Buffer Delay

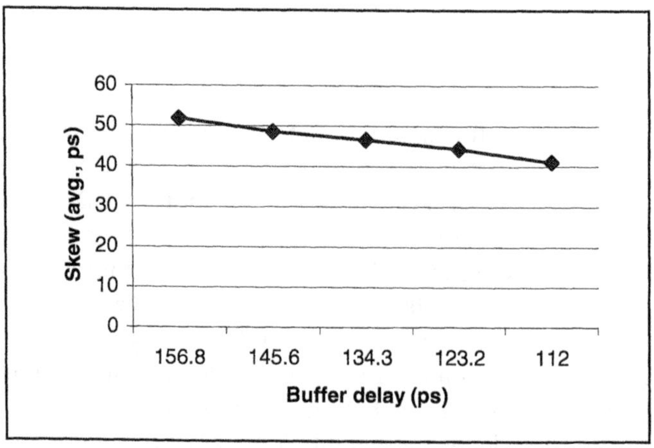

(b) Clock Skew vs. Buffer Delay

Figure 1-8. Sensitivity Simulation of Clock Buffers in a 0.13μm Process [49].

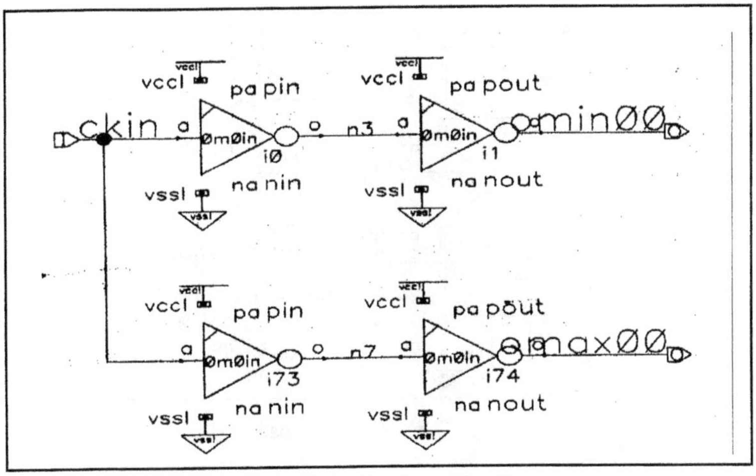

(a) CB buffer (heavy load)

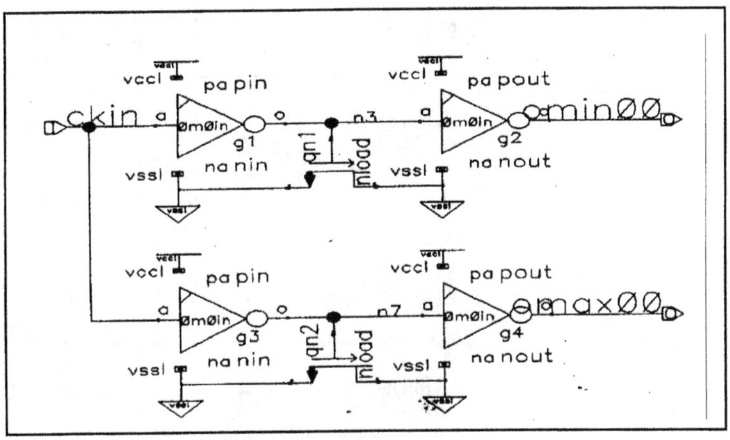

(b) CB Buffer (light load)

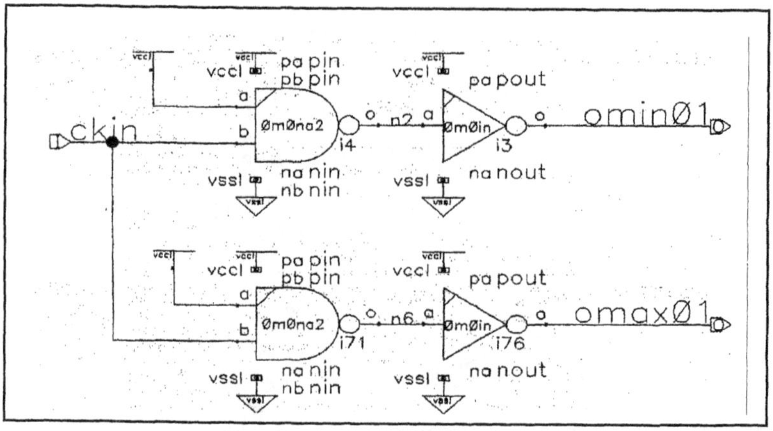

(c) SQ buffer (heavy load)

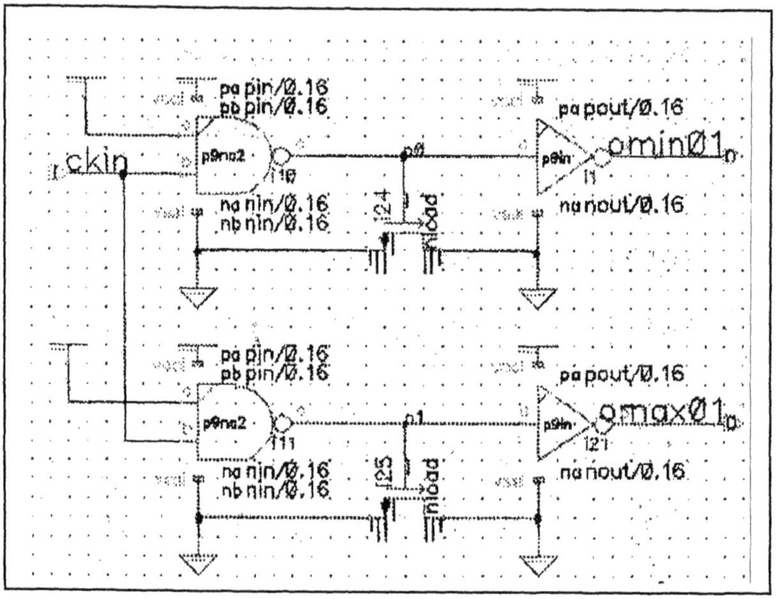

(d) SQ Buffer (light load)

Figure 1-9. Optimization Circuitry for Clock Buffer Sizing.

1.5 Power Supply and Reliability Issues

The power lines near large-size buffers should be protected by sufficient decoupling capacitors. The on-chip decoupling capacitors are usually formed with dummy transistors [57]. A conventional method is used to allocate on-chip decoupling capacitance based on the layout window rule – a percentage (i.e., 10%) of the layout area in each layout window. Closer to the large-size clock buffers, a high percentage (20% to 40%) of the layout area is reserved for the decoupling capacitors. The electromigration problem is another design concern for large-size clock buffers. The electromigration is the wear-out problem of metal lines when a DC current exists. Figure 1-10 shows the current DC paths that can cause the electromigrations if the current density of metal lines exceeds the required upper limits. The following limits for current densities of each metal layer are specified for a TSMC 0.18μm process in six metal layers [121].

- M1 < 1.0mA/μm
- M2 < 1.0mA/μm
- M3 < 1.0mA/μm
- M4 < 1.0mA/μm
- M5 < 1.0mA/μm
- M6 < 1.6mA/μm

The electromigration rules are also applied to the contacts and vias in the layout. The following limits for current densities of contact and vias are specified for the 0.18μm TSMC process in multiple layers of contact and vias [121].

- Contact < 0.53mA
- Via1 < 0.28mA
- Via2 < 0.28mA
- Via3 < 0.28mA
- Via4 < 0.28mA
- Via5 < 0.706mA

The above current density limits have to be followed in the clock buffer design for interconnect and via sizes. In addition, the AC current paths for large-size clock buffers need to prevent the "self-heating" problem, which is caused by the power consumption at localized high temperatures, burning down the metal lines. A set of RMS current density limits is also provided for a given process. In order to solve the electromigration and self-heating problems, the metal width is increased with more vias or contacts between the wider metals in two adjacent layers.

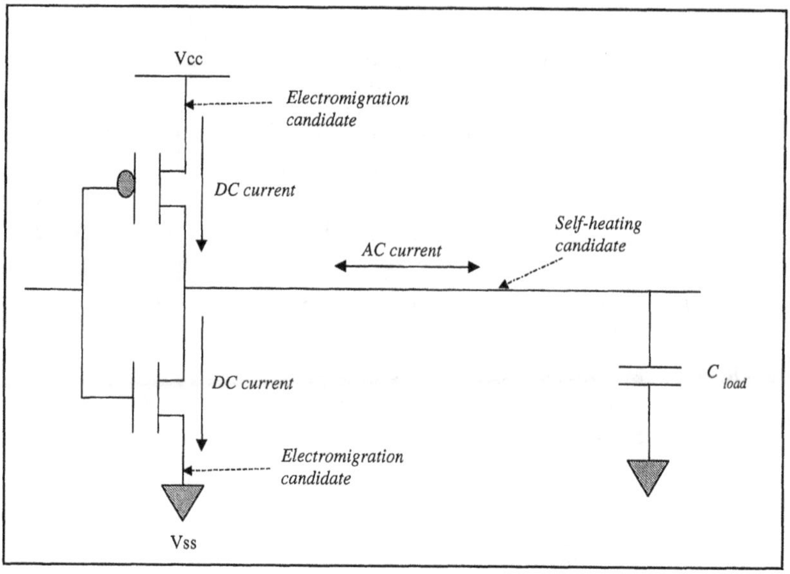

Figure 1-10. Potential Electromigration and Self-heating Failures in Clock Distribution.

1.6 Design Complexity of Clock Distribution

In a modern high-performance chip design, the clock signals can be in a complex form, which includes multiple clock frequencies and multiple clock phases, and even multiple clock pulse widths. Figure 1-11 shows the clock distribution hierarchy in an Intel microprocessor chip [58]. It consists of the three major blocks as follows: (a) the phase-locked loop (PLL) that synthesizes the chip clock from an off-chip system clock to on-chip clock; (b) the global clock distribution network that distributes the clock from the PLL to all functional blocks; and (c) the local clock generator that generates the local clock styles for the various functional blocks.

Figure 1-12(a) shows the different types of local clocks used in a microprocessor chip [58]. The local clocks consist of three different core clocks (slow, medium, and fast frequencies), two different clock phases (Phase 1 and 2) and two different styles (normal and pulse). Figure 1-12(b) shows the detailed diagram of the global clock distribution network [58]. The clock gating can occur at the main driver after the PLL or before the unit clock generator to shut down the clocks in order to save power, since the clock can consume up to 50% microprocessor full chip power. Delay compensator (delay loop line) is designed to skew or de-skew the local clocks relative to one another to further minimize the clock skew between the local clocks. The circuit designers will choose the right clock styles for local circuits. The appropriate local clock generators are used to produce the required clock signals. The following definitions are used for different core clock frequencies and clock phases. The timing phases for the normal and pulse clocks are shown in Figure 1-13.

- Reference clock. This is a medium frequency 50% duty cycle clock. All signal timings are referenced to this clock in the RTL codes.

- Normal clock. The rising and falling edges of this clock are generated from two separated edges of a reference clock. The pulse width of the normal clock is a function of the pulse width of the reference clock.

- Pulse clock. Both the rising and falling edges of this clock are generated from the same edge of the reference clock. The pulse width of the pulse clock is not a function of the pulse width or the frequency of the reference clock.

- Domino clock. Domino clocks are basically pulse clocks. These clocks are used to clock domino circuits and have minimum pulse width requirements.

- Static clock. These are the same as normal clocks and will be used interchangeably with them.

- Phase 1 clock. The rising edge of the clock is generated from the rising edge of the reference clock.

- Phase 2 clock. The rising edge of the clock is generated from the falling edge of the reference clock.

- Slow clock. 450MHz clock in 0.18μm process.

- Medium clock. 900MHz clock in 0.18μm process.

- Fast clock. 1.8GHz clock in 0.18μm process.

- Interface clocks. Special clocks are used for interfacing between two different frequency domains.

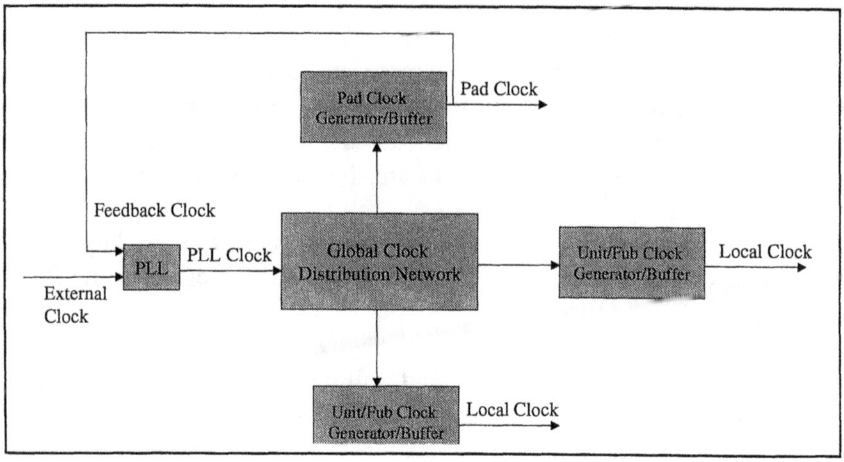

Figure 1-11. Clock Distribution Hierarchy in a Pentium-IV® Microprocessor [58].

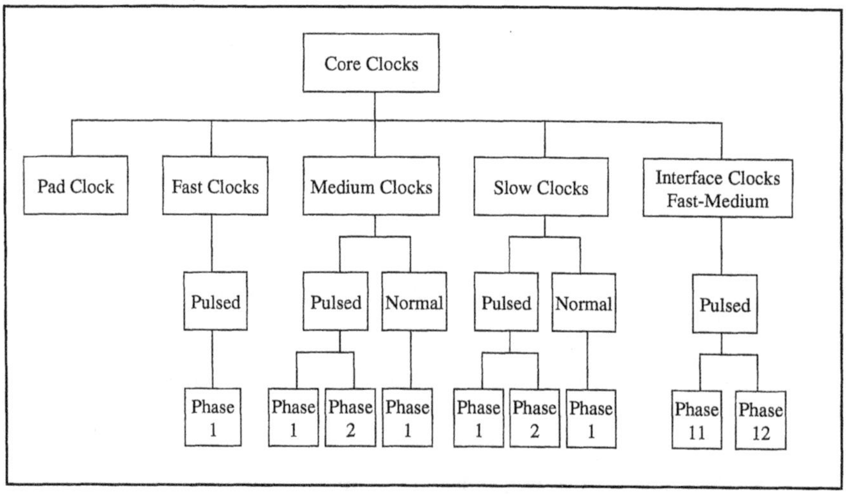

(a) Multiple Clock Phases.

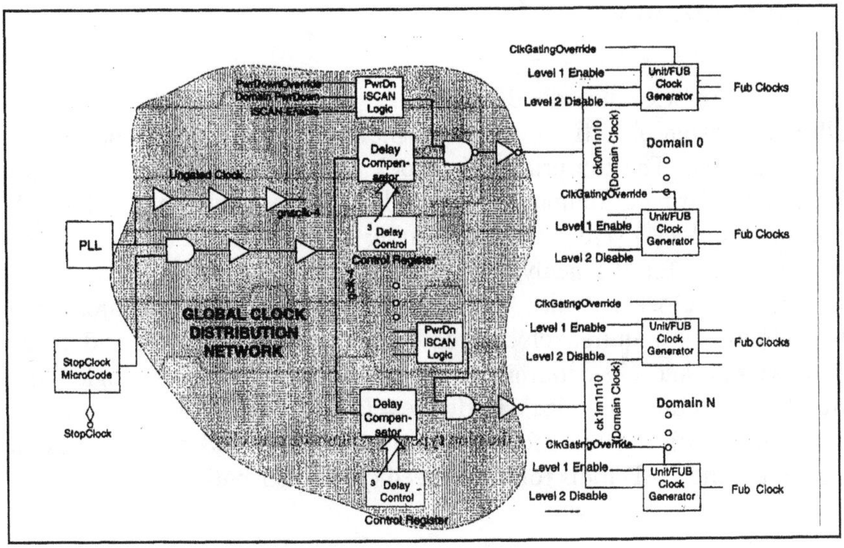

(b) Clock Distribution Network.

Figure 1-12. Clock Distribution for a Pentium-IV® Microprocessor [58].

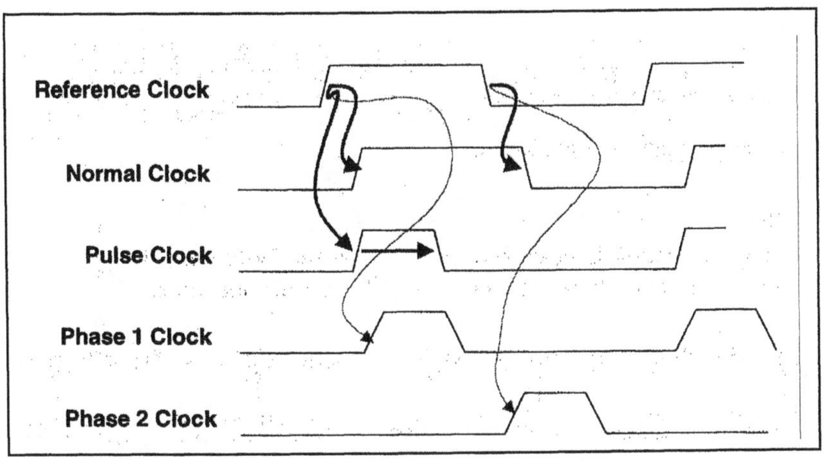

Figure 1-13. Clock Phases in an Intel Pentium-IV Microprocessor [58].

1.7 Summary

The clock networks design is the balance of three factors in the following design objectives: (a) clock skew, (b) power consumption, and (c) layout area. The top priority in clock network design is to minimize clock skew. A rule of thumb is that the clock skew should be less than 10-15% of the clock cycle time. The timing tools need the accurate clock skew numbers for the analysis of timing violations in the circuit. The tasks for the clock network design can be classified in two areas: (1) determining the required clock frequency, the clock phases, and clock pulse widths, and (2) determining the clock distribution topologies and circuitry, including the clock buffer levels and sizes. The focus of this book will mostly on the second area about circuit and physical design techniques and CAD tools for high-speed clock distribution.

Chapter 2 Overview to Timing Constraints

The sequential elements or storage elements in the digital circuits restore the data. The sequential elements include the flip-flop and latch circuits, which change the saved data based on the rising or falling clock pulses. The timing constraints of sequential elements (flip-flops and latches) define the minimum time intervals between data signals and clocks. They specify when signals must be ready in order to ensure the correct functioning of the sequential circuits. Timing constraints includes the following items: the setup time, the hold time, the recovery time, and the minimum pulse width. An example of a hold time problem arises from excessively long clock delay compared to the data delay.

2.1 Propagation Delay and Transition Time

The propagation delay through a gate is the sum of the gate intrinsic delay, the load-dependent delay, and the input-slew dependent delay. A delay is defined as the time interval between the input stimulus crossing 50% of Vdd and the output crossing 50% of Vdd. Figure 2-1 shows the propagation delay definition. Factors that affect the propagation delays include: temperature, supply voltage, and process variations, fanout loading, interconnect loading, input transition time, and input-signal polarity. The input transition time is defined as the time to go from 10%Vdd to 90%Vdd for the input signal. The output transition time is defined as the time to go from 10%Vdd to 90%Vdd for the output signal. In addition, for the signals shown in Figure 2-1, the transition time is further classified into two types as follows:

- Rising time for the waveform (signal) rising from Vss to Vdd.
- Fall time for the waveform (signal) falling from Vdd to Vss.

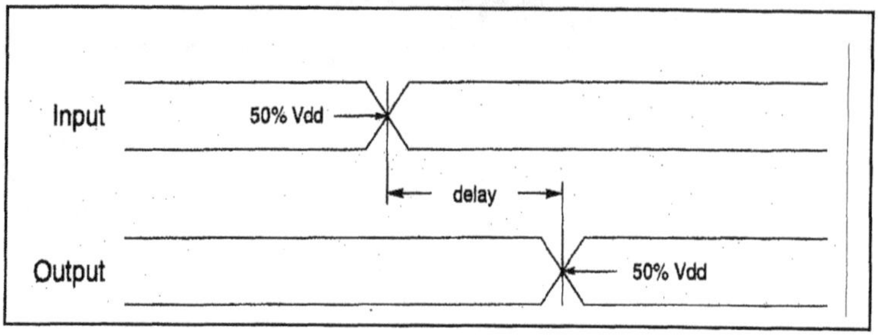

Figure 2-1. Propagation Delay.

2.2 Setup Time Constraint

The setup time for a sequential cell is the minimum length of the time that the data-input signal must remain stable prior to the active edge of the clock (or other specified signal) in order to ensure the correct functioning of the sequential element. The sequential circuit is considered functional as long as the delay for the output reaching its expected value does not exceed the reference delay (measured with a larger setup time) by more than 10%. Setup constraint value is measured as the interval between the data signal crossing 50% of Vdd and the clock signal crossing 50% of Vdd. For the measurement of setup time, the data input signal is kept stable after the active clock edge for a long hold time. Figure 2-2 illustrates the setup time definition for a positive edge-triggered sequential circuit.

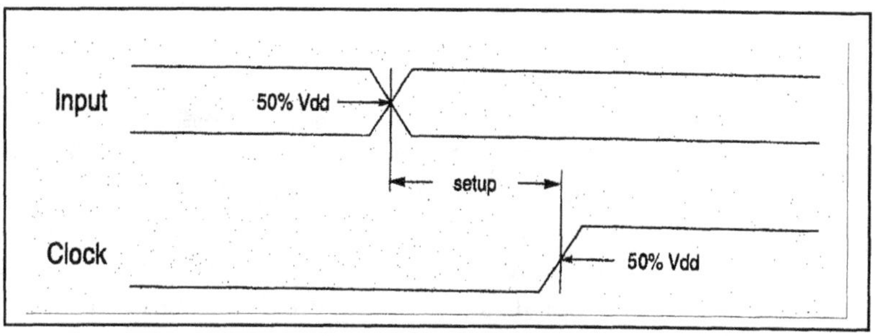

Figure 2-2. Setup Time.

The minimum requirements are simulated for the circuits of a flip-flop or latch. It avoids the meta-stability problem by maintaining sufficient setup and hold time for the input data bit at the clock rising edge (data capturing moment), since the flip flop is not ideal for capturing the right data immediately. The data must be stable enough long before the rising clock edge. Any setup time at the input data to the sequential element, that is less than the minimum requirement, is called a setup time violation. The setup time violations are caused by the delay differences in the clock paths and the data paths, since the setup time definition involves the relative difference between the data arriving edge and the clock arriving edge for the flip-flop or latch. To eliminate the setup time violation, one can either reduce the delays in the data paths or increase the delays in the clock paths. Increasing the clock delays has a global impact on all sequential elements. To reduce the delays of the data paths, the following methods are usually used.

- Increasing transistor sizes.
- Reducing buffer stages. But for long interconnects, adding repeaters in the middle will reduce RC delay.
- Reducing capacitive loads.

2.3 Hold Time Constraint

The hold time for a sequential element is the minimum length of time that the data input signal must remain stable after the active edge of the clock to ensure correct functioning of the circuit. The circuit is considered functional as long as the delay for the output reaching its expected value does not exceed the reference delay (measured with a large hold time) by more than 10%. The hold constraint value is measured as the interval between the clock signal crossing 50% of Vdd and the data signal crossing 50% of Vdd. For the measurement of hold time, the data input signal is held stable before the active clock edge for a long setup time. Figure 2-3 illustrates the hold time for a positive-edge-triggered sequential circuit.

The minimum hold time requirements are simulated for the circuits of flip flop or latch. Any hold time at the input data to the sequential element, that is less than the minimum requirement, is called a hold time violation. Just like setup time violations, hold time violations are also caused by the delay differences in the clock paths and the data paths, since the hold time definition involves the relative difference between the data arriving edge and the clock arriving edge at the flip flop or the latch. To eliminate hold time violations, one can either increase the delays in the data paths or reduce the delays in the clock paths. Reducing the clock delays has a global effect on all sequential elements. To increase the delays of the data paths, the following methods are usually used.

- Adding buffers.
- Reducing transistor sizes.
- Adding dummy loads.

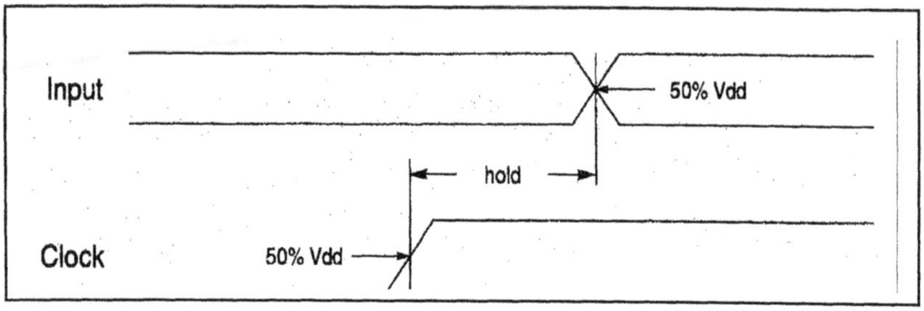

Figure 2-3. Hold Time.

2.4 Recovery Time and Pulse Width

The recovery time for sequential cells is the minimum length of time that the active-low set or reset signal must remain high before the active edge of the clock to ensure correct functioning of the sequential circuit. The circuit is considered functional as long as the delay for the output reaching its expected value does not exceed the reference delay (measured with a large recovery time) by more than 10%. The recovery constraint value is measured as the interval between the set or reset signal crossing 50% of Vdd and the clock signal crossing 50% of Vdd. For the measurement of recovery time, the set or reset signal is held stable after the active clock edge for a long hold time. Figure 2-4 illustrates the definition of recovery time.

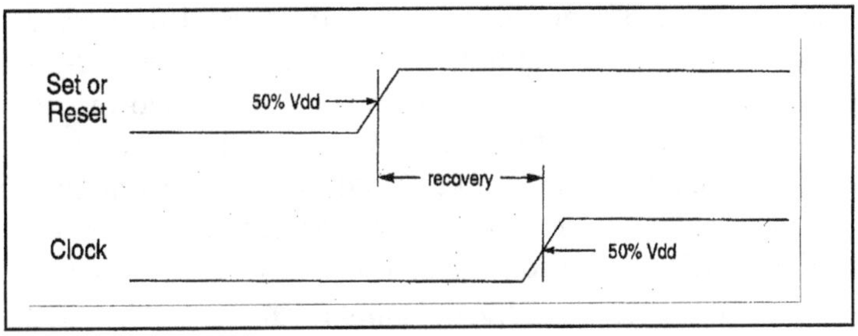

Figure 2-4. Recovery Time.

Minimum pulse width is the minimum length of time between the leading and trailing edges of a pulse waveform. Minimum pulse width high (minpwh) is the interval between the rising edge of the signal crossing 50% of Vdd and the falling edge of the signal crossing 50% of Vdd. Minimum pulse width low (minpwl) is measured as the interval between the falling edge of the signal crossing 50% of Vdd and the rising edge of the signal crossing 50% of Vdd. Figure 2-5 illustrates these definitions of minimum pulse width. The minimum pulse width is also defined for the clock signal as well as the set and reset signals (SN, RN). An input pulse of shorter duration will produce unpredictable results.

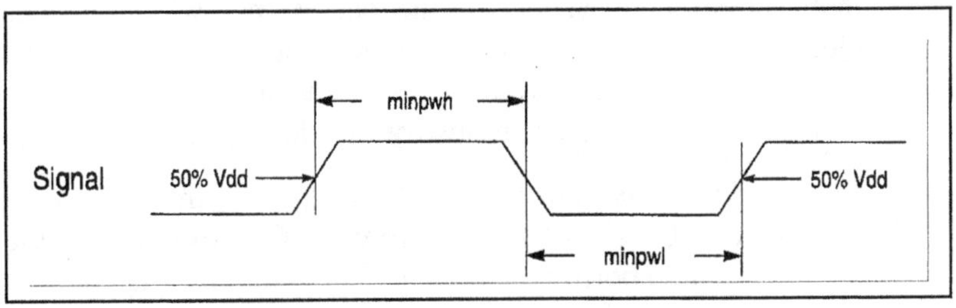

Figure 2-5. Minimum Pulse Width.

2.5 Time Borrowing

Time borrowing is a powerful technique in high-speed digital circuit design. It relaxes the sensitivity to clock skew and jitter by requiring less margin on the average. A circuit is said to be time borrowing if it uses time from the previous (or next) phase (or cycle) to perform its logic. A common example of time borrowing circuit is a two-phase latch circuit. If one phase can produce valid data early, the next phase can utilize the unused time from this phase. Figure 2-6 shows this concept, where phase 1 has very short logic delay and the unused time from phase 1 is used in the phase 2's logic. Time borrowing can be deliberate where one phase (or cycle) is designed longer at the expense of the previous or next cycle. Time borrowing can also be opportunistic or unintentional where occasionally data may arrive late because of clock jitter or supply voltage and signal coupling variation etc. and can borrow into the next phase (or cycle). Data can arrive early, too, to provide a head start for the next phase (cycle) logic due to time borrowing. Time borrowing allows the phase (cycle) delays to be averaged out over the entire path. There are time borrowing opportunities as follows in the design [58]:

- *Backward time borrowing.* This type of circuit allows the time borrowing from the previous phase or cycle only. Figure 2-6(a) shows the time borrowing window for a two-phase static latch. The potential time borrowing window is equal to the clock pulse width minus the latch delay as shown in the figure.
- *Forward time borrowing.* This type of circuit allows time borrowing from the next phase or cycle only. Figure 2-6(b) shows the time borrowing window for a domino. In order for the domino stage to evaluate properly, there is a minimum pulse width intersection requirement for the data signal with the evaluate clock. This leaves the rest of the clock pulse width available for borrowing by the previous phase. The actual time borrowing window is even smaller because of the clock inaccuracy.
- *Bi-directional time borrowing.* This type of circuit allows both forward and backward time borrowing. One example of the bi-directional time borrowing circuit is a set dominant latch circuit.

The time borrowing window is the window when an input signal may propagate to the next phase logic without incurring more than the minimum possible delay. For a latch or a domino circuit, this window starts just after the rise of the clock signal. Arriving sooner than this will result in the clock to determine the output's valid time. The time borrowing window ends at the setup time. Arriving later than the setup time will result in delay push-out and failure. A signal arriving too soon and "waiting" is a waste of resources; that path might have been downsized or could have taken some more logic. The wider this time borrowing window, the bigger is the available time borrowing. Master slave flip-flops are edge sensitive – it has practically zero time borrowing window. For an MSFF to another MSFF cycle path, the amount of useful logic that can be placed is calculated as follows:

$$T_{\log ic} = T_{cycle} - T_{setup} - T_{clk\_out} - T_{skew}\ . \qquad (2\text{-}1)$$

Here T_{logic} is the maximum delay allowed for the logic between two flip-flops. T_{cycle} is the cycle time; T_{setup} the setup time to the flip-flop; T_{clk_out} the clock to output delay, and T_{skew} the clock skew and clock jitter margins. For example, using some typical values for the various times, $T_{logic} = 1111\text{-}40\text{-}135\text{-}122 = 814ps$. The useful logic delay is only 73% of the cycle time and the remaining 27% is wasted.

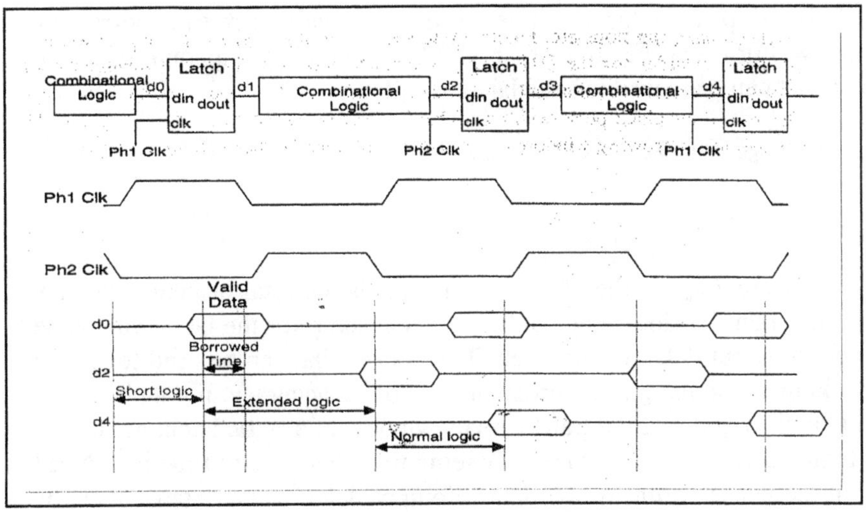

(a). Two Phase Static Circuit.

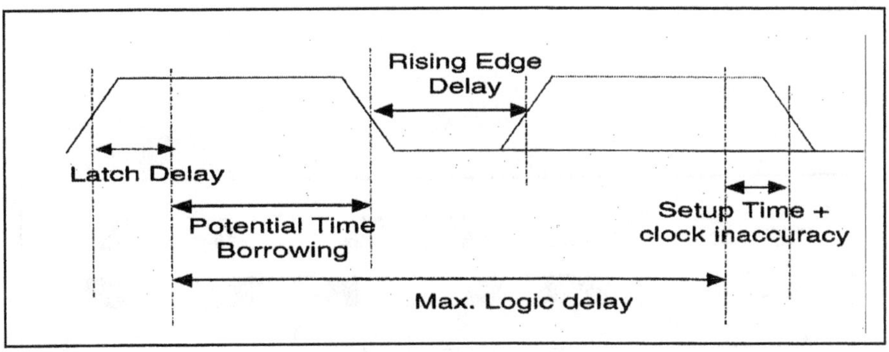

(b) Domino Circuit

Figure 2-6. Time Borrowing Concepts [58].

The time borrowing window is the window when an input signal may propagate to the next phase logic without incurring more than the minimum possible delay. For a latch or a domino circuit, this window starts just after the rise of the clock signal. Arriving sooner than this will result in the clock to determine the output's valid time. The time borrowing window ends at the setup time. Arriving later than the setup time will result in delay push-out and failure. A signal arriving too soon and "waiting" is a waste of resources; that path might have been downsized or could have taken some more logic. The wider this time borrowing window, the bigger is the available time borrowing. Master slave flip-flops are edge sensitive – it has practically zero time borrowing window. For an MSFF to another MSFF cycle path, the amount of useful logic that can be placed is calculated as follows:

$$T_{\log ic} = T_{phase} - T_{setup} - T_{skew}. \quad (2\text{-}2)$$

Here T_{phase} is the phase width, T_{skew} is the total inaccuracy (skew and jitter) on the clock edge and T_{setup} is the setup for the latch with some clock to output delay of the latch. Time borrowing can be used to reduce or eliminate the margin requirement for clock inaccuracy. If data becomes valid at the middle of the transparency window of a static latch, no margin is required for clock inaccuracy or setup time. The useful time is reduced from the phase width by the data to output delay of the latch only. The data flows transparently through the latches, unencumbered by the clock, as shown in Figure 2-7. In order to centre the valid time of the data in the transparency window of the latch, a big margin is required in the first stage.

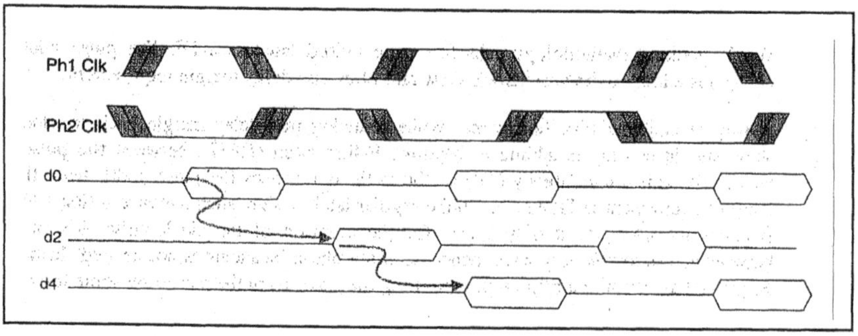

Figure 2-7. Time Borrowing To Avoid Clock Inaccuracy.

2.6 One Example: Fixing Hold Time Violations

Figure 2-8 shows an example for the hold time violation and engineering change order (ECO) technique to repair the hold time violation [60]. The ext_clk is the chip pin connected with the external clock, and ext_data is the chip pin connected with the external data. The ext_clk and ext_data have to meet a minimum hold time requirement in order to make the flip-flop DFFHQX1 work properly. In other words, the minimum hold time for ext_clk and ext_data depends on the delay differences from the ext_clk pin to the clock input of the flip-flop DFFHQX1 and from the pin ext_data to the data input of the flip-flop DFFHQX1. Figure 2-9(a) shows the minimum hold time requirement for ext_clk and ext_data in the circuit shown in Figure 2-8. This will satisfy the minimum hold time requirement, as shown in Figure 2-9(b), for the clock input (clk) and data input (d) for DFFHQX1, where q_1 is the output from DFFHQX1.

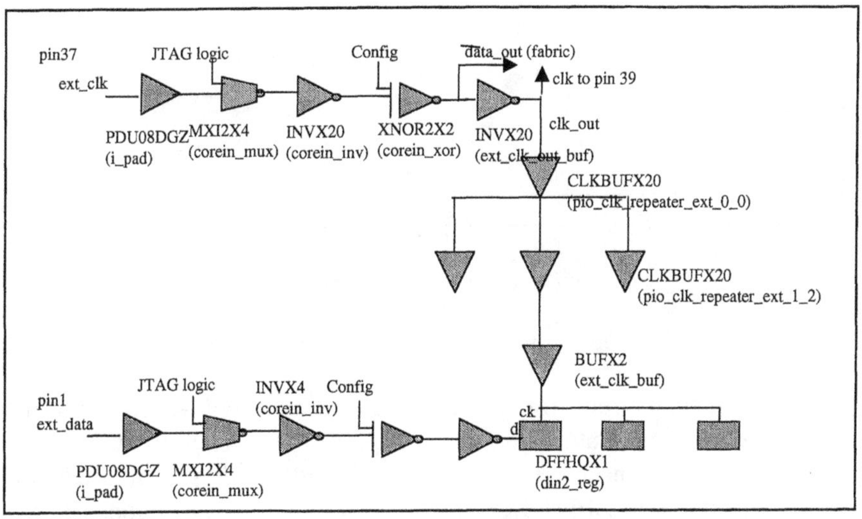

Figure 2-8. Circuit Description for Hold Time Violation Example [60].

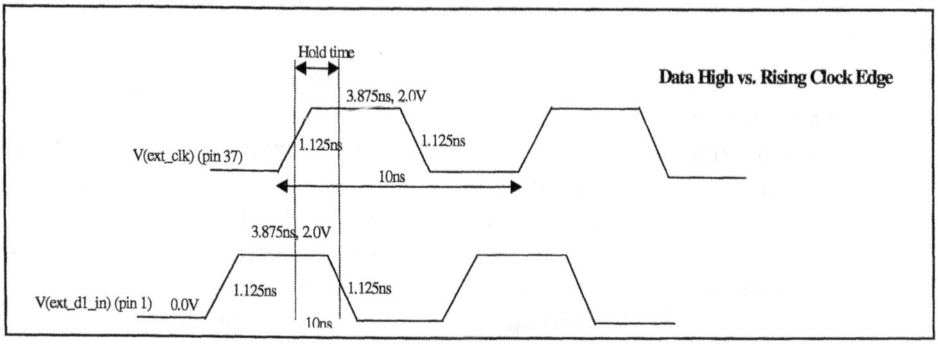

(a) Ext_clk vs. Ext_data at Inputs to Chip Pins (Pin 37 and Pin 1).

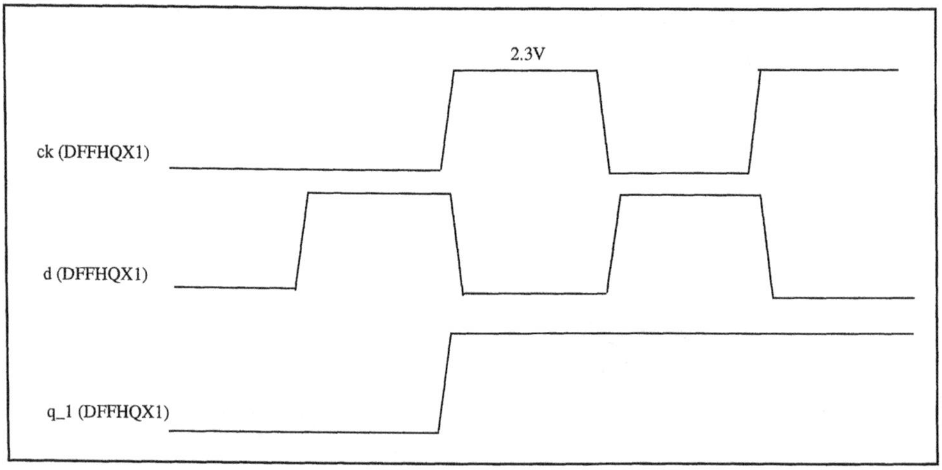

(b) ck vs. d at Inputs to Flip-flop (DFFHQX1).

Figure 2-9. Minimum Hold Time Requirements [60].

A new circuit for reducing the minimum hold-time requirement of ext_clk vs. ext_data is shown in Figure 2-10. The circuit reduces four inversions in the original clock distribution path shown in Figure 2-8, from ext_clk to DFFHQX1. It also adds two more inversions (BUFX2) in the data path from the ext_data to DFFHQX1. The minimum hold-time requirement at chip pins is reduced due to our speeding up for the clock path and meanwhile slowing down the data path by two inverters. Here, no new logic gates are added in the circuit and only interconnections of the gates are changed to reduce the hold-time margin between ext_clk and ext_data. All the logic functions are reserved in the new circuit and little delay impact for other parts of the chip.

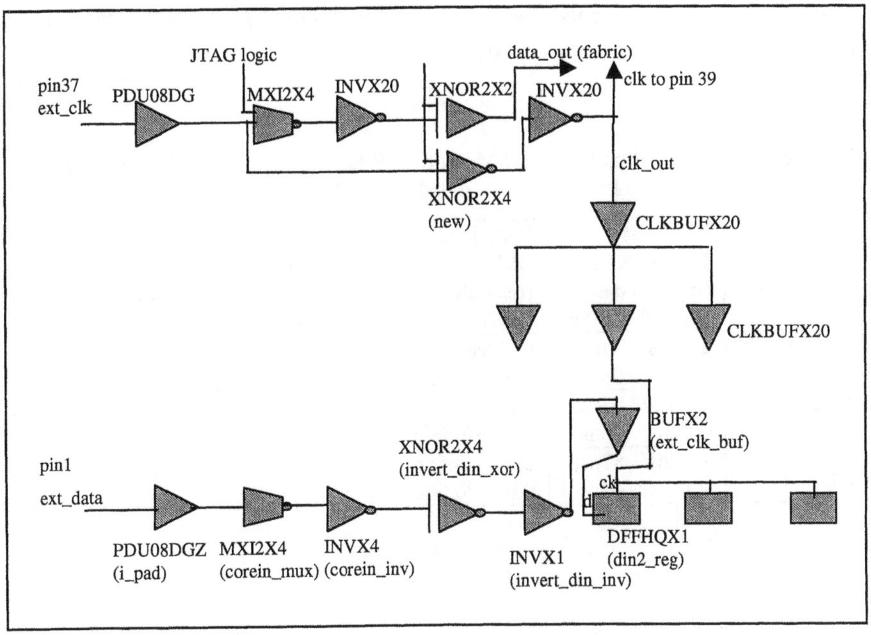

Figure 2-10. New Circuit Changed to the Circuit Shown in Figure 2-9 [60].

Table 2-1 shows the minimum hold time data simulated in a 0.25μm process. The simulation is done in the slow process corner for the old circuit as shown in Figure 2-8. The mismatch between the simulation and measurement results are about 1%-13% in various core and IO supply voltages. Table 2-2 shows the minimum hold time results in the simulation and measurement at the fast process corner for the circuit shown in Figure 2-8. The minimum hold time is more than 1.0ns, which is beyond the external design specification [60]. The new circuit, as shown in Figure 2-11, reduces the minimum hold time requirement between ext_data and ext_clk, and the simulation results in various process corners for the 0.25μm TSMC process are shown in Tables 2-3 to 2-6 at 85°C and 0°C temperatures in slow and fast process corners.

Figure 2-11 shows the simulation waveforms for the minimum hold times between the input signals ext_d1_in and ext_cin. Using the new circuit shown in Figure 2-10 compared to the old circuit shown in Figure 2-8, the minimum hold time is reduced to about 0.4ns. This example shows that the hold time violations can be fixed by the following approaches: (a) speeding up the clock or reducing the delays in the clock distribution paths, (b) slowing down the data or increasing the delays in the data distribution paths, or a combinations of (a) and (b). Note that the hold time violations must be fixed in the design, since they cannot be eliminated by a lower clock frequency.

Table 2-1. Hold Time Simulation and Measurement Results for Circuit Shown in Figure 2-8. Slow Process Corner in 0.25μm TSMC Process, 85°C.

Core and IO Voltages	Clk_in at DFFHQXZ1 (rise/fall delay, rise/fall time)	Data_in at DFFHQX1 (rise/fall delay, rise/fall time)	Minimum hold time (ext_data vs. ext_clk)	Tester's hold time measurement. Mismatch with simulation (%)
2.25V/3.0 V	2.24ns/2.23ns, 0.13ns/0.08ns	0.78ns/0.72ns, 0.08ns/0.05ns	1.60ns	1.66ns, 4%
2.50V/3.3 V	2.08ns/2.02ns, 0.12ns/0.08ns	0.75ns/0.64ns, 0.07ns/0.04ns	1.70ns	1.96ns, 13%
2.75V/3.6 V	1.98ns/1.87ns, 0.11ns/0.07ns	0.74ns/0.58ns, 0.07ns/0.04ns	1.90ns	1.91ns, 1%

Table 2-2. Hold Time Simulation and Measurement Results for Circuit Shown in Figure 2-8. Fast Process Corner in 0.25□m TSMC Process, 85°C.

Core and IO Voltages	Clk_in at DFFHQXZ1 (rise/fall delay, rise/fall time)	Data_in at DFFHQX1 (rise/fall delay, rise/fall time)	Minimum hold time (ext_data vs. ext_clk)	Tester's hold time measurement. Mismatch with simulation (%)
2.25V/3.0 V	1.55ns/1.42ns, 0.09ns/0.06ns	0.60ns/0.44ns, 0.05ns/0.04ns	1.20ns	1.43ns, 16%
2.50V/3.3 V	1.50ns/1.32ns, 0.08ns/0.06ns	0.61ns/0.39ns, 0.05ns/0.03ns	1.50ns	1.84ns, 18%
2.75V/3.6 V	1.47ns/1.23ns, 0.08ns/0.05ns	0.62ns/0.34ns, 0.05ns/0.03ns	1.70ns	1.86ns, 3%

Table 2-3. Hold Time Simulation Results for Circuit Shown in Figure 2-10. Slow Process Corner in 0.25μm TSMC Process, 85°C.

Core and IO Voltages	Clk_in at DFFHQXZ1 (rise/fall delay, rise/fall time)	Data_in at DFFHQX1 (rise/fall delay, rise/fall time)	Minimum hold time (ext_data vs. ext_clk) (ext_data is high/low worst)
2.25V/3.0V	1.67ns/1.60ns, 0.22ns/0.19ns	1.24ns/1.11ns, 0.12ns/0.09ns	0.39ns
2.50V/3.3V	1.55ns/1.46ns, 0.21ns/0.18ns	1.14ns/1.01ns, 0.11ns/0.07ns	0.39ns
2.75V/3.6V	1.46ns/1.37ns, 0.19ns/0.17ns	1.07ns/0.94ns, 0.10ns/0.07ns	0.39ns

Table 2-4. Hold Time Simulation Results for Circuit Shown in Figure 2-10. Fast Process Corner in 0.25μm TSMC Process, 85°C.

Core and IO Voltages	Clk_in at DFFHQXZ1 (rise/fall delay, rise/fall time)	Data_in at DFFHQX1 (rise/fall delay, rise/fall time)	Minimum hold time (ext_data vs. ext_clk) (ext_data is high/low worst)
2.25V/3.0V	1.18ns/1.08ns, 0.16ns/0.15ns	0.86ns/0.71ns, 0.08ns/0.06ns	0.29ns
2.50V/3.3V	1.14ns/1.02ns, 0.15ns/0.14ns	0.82ns/0.67ns, 0.08ns/0.05ns	0.29ns
2.75V/3.6V	1.10ns/0.97ns, 0.15ns/0.14ns	0.79ns/0.63ns, 0.07ns/0.05ns	0.29ns

Table 2-5. Hold Time Simulation Results for Circuit Shown in Figure 2-10. Slow Process Corner in 0.25μm TSMC Process, 0°C.

Core and IO Voltages	Clk_in at DFFHQXZ1 (rise/fall delay, rise/fall time)	Data_in at DFFHQX1 (rise/fall delay, rise/fall time)	Minimum hold time (ext_data vs. ext_clk)
2.25V/3.0V	1.49ns/1.44ns, 0.19ns/0.16ns	1.10ns/0.99ns, 0.10ns/0.07ns	0.29ns
2.50V/3.3V	1.38ns/1.33ns, 0.18ns/0.15ns	1.00ns/0.90ns, 0.09ns/0.06ns	0.29ns
2.75V/3.6V	1.31ns/1.24ns, 0.17ns/0.15ns	0.95ns/0.84ns, 0.09ns/0.06ns	0.29ns

Table 2-6. Hold Time Simulation Results for Circuit Shown in Figure 2-10. Fast Process Corner in 0.25μm TSMC Process, 0°C.

Core and IO Voltages	Clk_in at DFFHQXZ1 (rise/fall delay, rise/fall time)	Data_in at DFFHQX1 (rise/fall delay, rise/fall time)	Minimum hold time (ext_data vs. ext_clk)
2.25V/3.0V	1.08ns/0.99ns, 0.15ns/0.13ns	0.78ns/0.66ns, 0.07ns/0.05ns	0.29ns
2.50V/3.3V	1.04ns/0.93ns, 0.14ns/0.13ns	0.75ns/0.61ns, 0.07ns/0.05ns	0.29ns
2.75V/3.6V	1.01ns/0.89ns, 0.14ns/0.12ns	0.72ns/0.58ns, 0.06ns/0.05ns	0.29ns

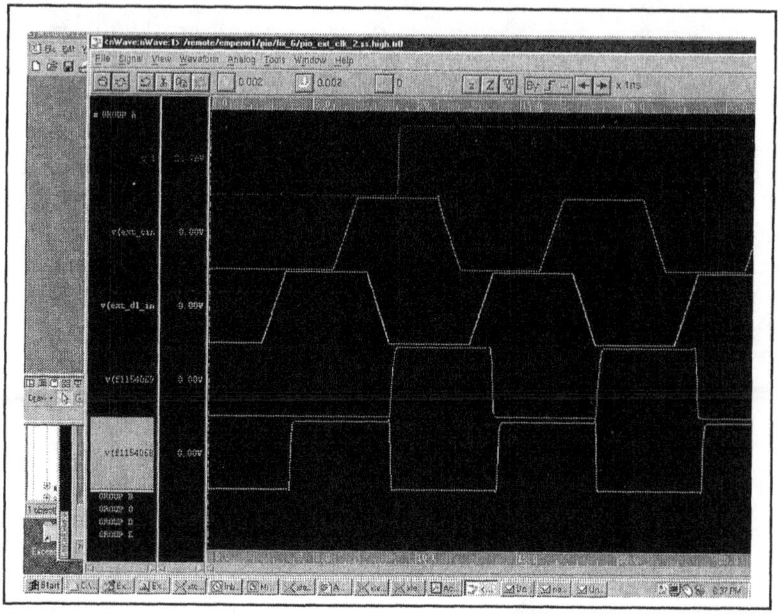

(a) Old Circuit

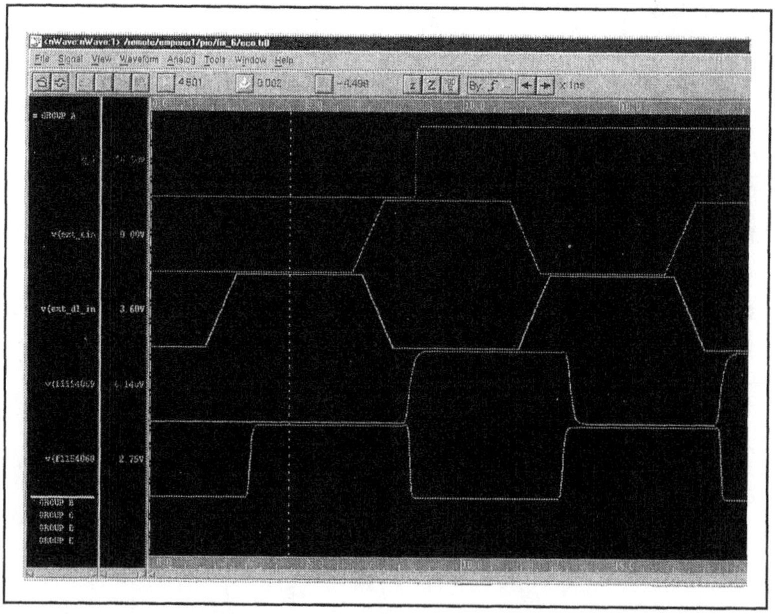

(b) New Circuit

Figure 2-11. Waveforms for Measuring the Minimum Hold Time Between ext_clk and ext_data [60].

2.7 Delay Slack Graph for Timing Constraints

One way to describe the timing constraints in a digital circuit is through a delay slack graph. The delay slack graph is directed from the primary inputs to primary outputs for the circuit. The number of arrival times and required arrival times needed to completely constrain the timing of a circuit is proportional to the number of sequential logic elements. On the other hand, the number of path-based constraints has been observed to grow exponentially with circuit size. The delay slack graph is a directed graph where each pin in the netlist is represented by a node, and each signal path between pins by an edge. For example, the timing of the circuit in Figure 2-12(a) can be represented by the delay slack graph as shown in Figure 2-12(b) [61]. The start and end points of the free edges are determined by the netlist, and the internal edges determined by the timing arcs defined in the library. Each node in the graph has an associated timing slack, T_{slack}. A positive slack means that the node is earlier than the timing constraint. A negative slack means that the node is late. Using the setup and hold times for flip-flops and latches, target clock cycle time, and clock skew, the slack graph generator can automatically derive the arrival time T_{arr} at the flip-flop clock inputs, and the required arrival time T_{reqArr} at the flip-flop data inputs. The T_{arr} for the primary inputs and T_{reqArr} for primary outputs have to be specified by users.

The slack is computed in two steps. First, the graph is traversed forward starting from T_{arr}, calculating the delay to each node. Then the graph is traversed backward starting from T_{reqArr}, calculating the required arrival time at each node. The difference is Tslack. Figure 2-13(a) shows the delays associated with each edge of the slack graph. The graph is traversed forward starting from T_{arr}, calculating the delay to each node. When two paths merge, the larger delay is used as shown in Figure 2-13(b). Next, the graph is traversed backward starting from T_{reqArr}, calculating the required arrival time at each node. When two paths merge, the smaller delay is used, as illustrated in Figure 2-13(c). The difference between the delay and required arrival time on each node is T_{slack}, as shown in Figure 2-13(d). The timing driven placement and in-placement logic optimization algorithms recognize that the path with negative slack needs improvement.

2.8 Summary

The setup time and hold time constraints for latches and flip-flops in the digital design have been discussed in this chapter. Both hold time and setup time requirements are racing conditions between clock signal and data signal arriving at these sequential elements. We show an example of how to reduce hold time violations by making interconnection changes in the clock distribution tree and adjusting the delays of the clock path and the data path. A delay slack graph is commonly used to represent the timing constraints in the timing driven placement and routing tools and static timing analysis tools [61,62].

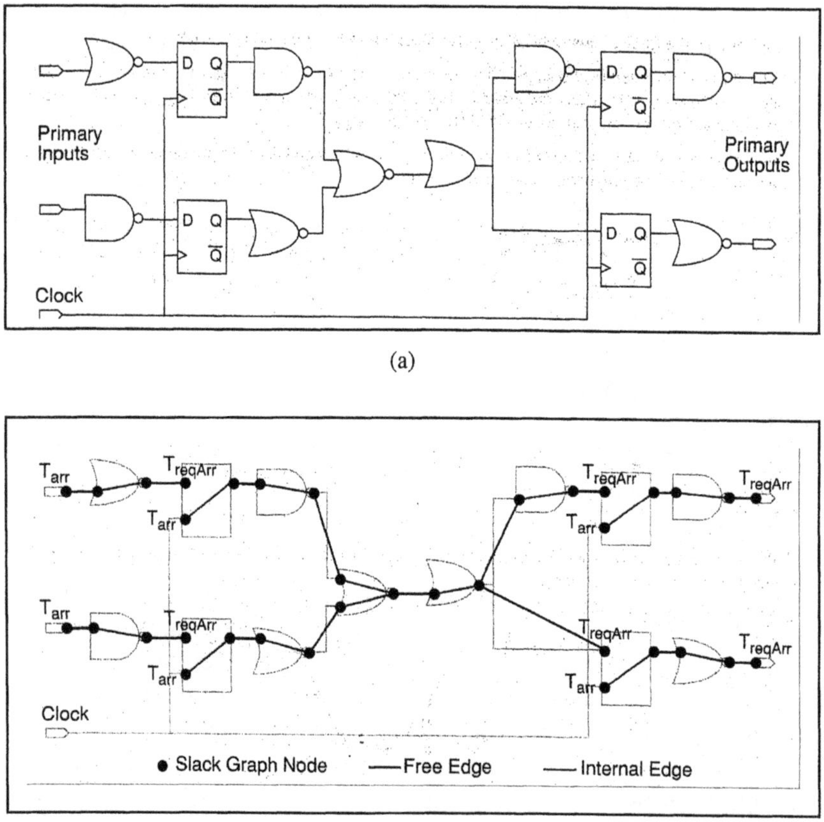

Figure 2-12. Delay Slack Graph [61].

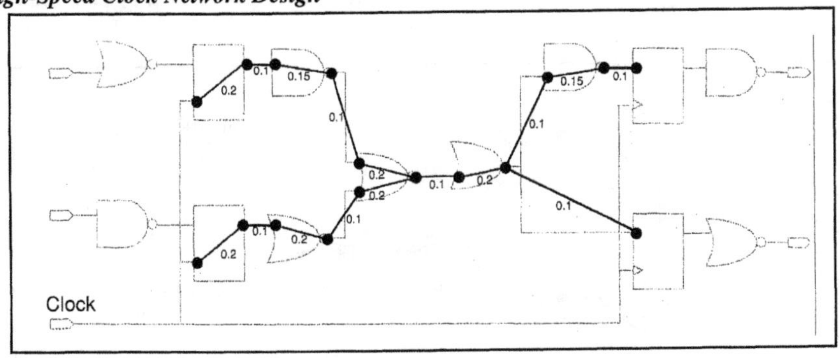
0.2
0.1
0.15
0.1
0.2
0.2
0.1
0.2
0.1
0.1
0.2
0.1
0.2
0.15
0.1
0.1
0.1
Clock

(a)

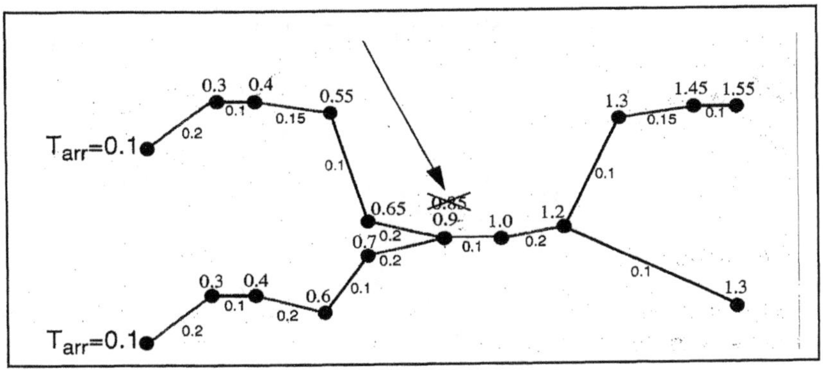
Tarr=0.1
0.2
0.3
0.1
0.4
0.15
0.55
0.1
0.65
0.85
0.9
0.7
0.2
0.2
0.1
1.0
0.2
1.2
0.1
1.3
0.15
1.45
0.1
1.55
0.1
1.3
Tarr=0.1
0.2
0.3
0.1
0.4
0.2
0.6
0.1

(b)

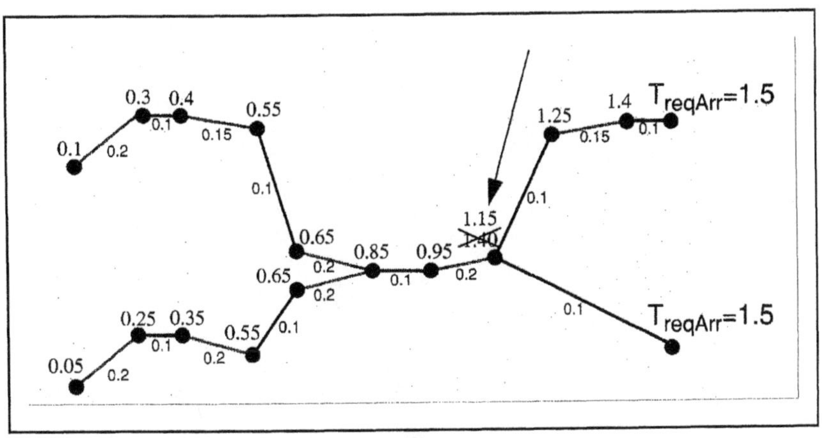
0.1
0.2
0.3
0.1
0.4
0.15
0.55
0.1
0.65
0.2
0.85
0.1
0.95
0.2
1.15
1.40
0.1
1.25
0.15
1.4
0.1
TreqArr=1.5
0.1
TreqArr=1.5
0.65
0.2
0.05
0.2
0.25
0.1
0.35
0.2
0.55
0.1

(c)

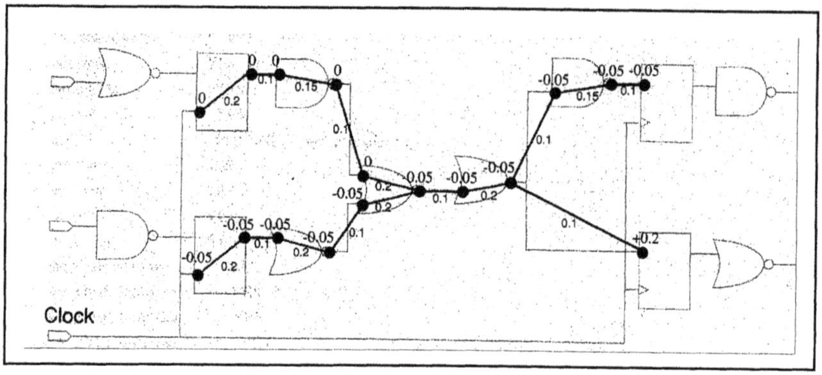

(d)

Figure 2-13. Slack Computation Steps [61].

Chapter 3 Sequential Clocked Elements

The sequential clocked elements provide the basic data storage elements in a sequential digital circuit. The sequential clocked elements include latch, flip-flop, and other storage cells. The sequential machine is a machine in which all events occur in a timely order. In order to determine the state of the machine, we need to reference the rising or falling edge of the clock signal to transfer the data in the machine. The storage circuits, such as the latch, flip-flop, or storage elements, are used to hold and transfer the data based on the clock rising and falling edges or high and low levels. Section 3.1 describes the latch circuits. Section 3.2 shows the flip-flop circuits. Section 3.3 discusses techniques to reduce the power consumption for sequential logic. Section 3.4 provides the summary to this chapter.

3.1 Latch Clocking

A latch is in transparent mode or open mode when the clock is asserted, and input data will propagate to the output. In the open mode, the latch output will change as the input data changes. The latch is in the opaque mode or close mode when the clock is de-asserted. In the close mode, it will disconnect the input data from the output, and the latch output will remain at its previous value. Figure 3-1 shows the circuit schematic for a clock high-level sensitive latch circuit [65]. The working principles of this latch are as follows.

- When the clock clk is high, it will turn on the pass-gate i69, and the latch is in the open mode. The input data d can pass through the pass gate i69 and get two inversions by i78 and the last stage pout or nout to the output o.

- When the clock clk is low, it will turn off the pass-gate i69 and block the input data from connecting to the output. The latch output remains at its previous value by the feedback inverter i73, and the latch is in the close mode.

- In order to avoid the contention between the new data and previous data while the latch is open, the tri-state inverter i77 is used to cut off the feedback path.

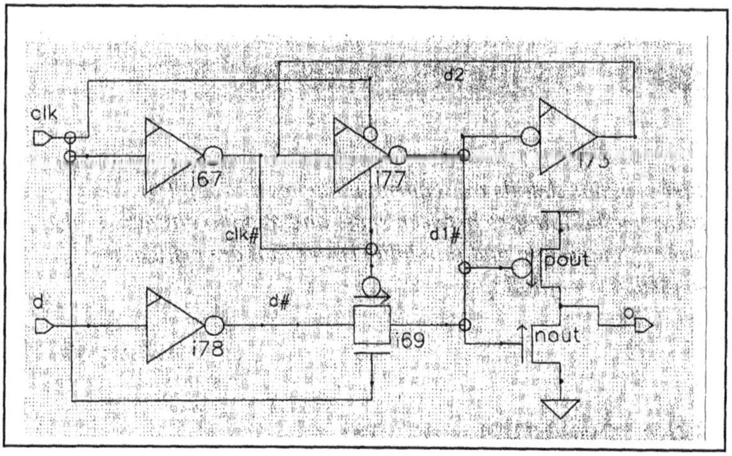

Figure 3-1. Latch Circuit [65].

In the following sections, we describe the variants from the basic latch circuit in a high-speed standard cell library [59]. Table 3-1 shows the list of latches and flip-flops provided from this standard cell library. They can be triggered by either a low or high clock edge, and may have scan and reset capabilities. The various latches and flip-flops provided in this cell library improve logic functions and testability for the design. Figure 3-2 shows the schematic of an active-low D-type transparent latch with asynchronous active-low reset (RN) and enable signal (GN). When the enable (GN) is low, the data (D) is transferred to the output (Q, QN). The truth table is shown in Table 3-2. Figure 3-3 shows an active-low D-type transparent latch with asynchronous active-low set (SN) and reset (RN). The set (SN) will dominate over the reset (RN) signal. When the enable (GN) is low, data (D) is transferred to the outputs (Q, QN). The truth table for this latch is shown in Table 3-3. Figure 3-4 shows an active-high D-type transparent latch with active-high output enable (OE). When the enable (G) is high and the output enable (OE) is high, the data (D) is transferred to the output (Q). Table 3-4 shows the truth function.

Table 3-1. Flip-Flops and Latches in a 0.25μm Standard Cell Library [59].

Cell	Logic Function
DFF	The DFF cell is a positive-clock-edge triggered, static D-type flip-flop.
DFFHQ	The DFFHQ cell is a high-speed, positive-clock-edge triggered, static D-type flip-flop. The cell has a single output (Q) and fast clock-to-out path.
DFFN	The DFFN is a negative-clock-edge triggered, static D-type flip-flop.
DFFNR	The DFFNR is a negative-clock-edge triggered, static D-type flop-flop with asynchronous active-low reset (RN).
DFFNS	The DFFNS is a negative-clock-edge triggered, static D-type flip-flop with asynchronous active-low set (SN).
DFFNSR	The DFFNSR is a negative-clock-edge triggered, static D-type flip-flop with asynchronous active-low reset (RN) and set (SN), and set dominating reset.
DFFR	The DFFR is a positive-clock-edge triggered, static D-type flip-flop with asynchronous active-low reset (RN).
DFFRHQ	The DFFRHQ is a high-speed, positive-clock-edge triggered, static D-type flip-flop with asynchronous active-low reset (RN). The cell has a single output (Q) and fast clock-to-out path.
DFFS	The DFFS is a positive-clock-edge triggered, static D-type flip-flop with asynchronous active-low set (SN).
DFFSHQ	The DFFSHQ is a high-speed, positive-clock-edge triggered, static D-type flip-flop with asynchronous active-low set (SN). The cell has a single output (Q) and fast clock-to-out path.
DFFSR	The DFFSR is a positive-clock-edge triggered, static D-type flip-flop with asynchronous active-low reset (RN) and set (SN), and set dominating reset.
DFFSRHQ	The DFFSRHQ is a high-speed, positive-clock-edge triggered, static D-type flip-flop with asynchronous active-low reset (RN) and set (SN), and set dominating reset. The cell has a single output (Q) and fast clock-to-out path.
DFFTR	The DFFTR is a positive-clock-edge triggered, static D-type flip-flop with synchronous active-low reset (RN).
JKFF	The JKFF is a positive-clock-edge triggered JK-type flip-flop.
JKFFR	The JKFFR is a positive-clock-edge triggered JK-type flip-flop with asynchronous active-low reset (RN).
JKFFS	The JKFFS is a positive-clock-edge triggered JK-type flip-flop with asynchronous active-low set (SN).
JKFFSR	The JKFFSR is a positive-clock-edge triggered JK-type flip-flop with asynchronous active-low reset (RN) and set (SN), and set dominating reset.
RSLAT	The RSLAT is an RS-type latch with active-high set (S) and reset (R).
RSLATN	The RSLATN is an RS-type latch with active-low set (SN) and reset (RN).
SDFF	The SDFF is a positive-clock-edge triggered, static D-type flip-flop with scan input (SI) and active-high scan enable (SE).
SDFFHQ	The SDFFHQ is a positive-clock-edge triggered, static D-type flip-flop with scan input (SI) and active-high scan enable (SE). The cell has a single output (Q) and fast clock-to-out path.
SDFFN	The SDFFN is a negative-clock-edge triggered, static D-type flip-flop with scan input (SI) and active-high scan enable (SE).
SDFFNR	The SDFFNR is a negative-clock-edge triggered, static D-type flip-flop with scan input (SI), active-high scan enable (SE), and asynchronous

	active-low reset (RN).
SDFFNS	The SDFFNS is a negative-clock-edge triggered, static D-type flip-flop with scan input (SI), active-high scan enable (SE), and asynchronous active-low set (SN).
SDFFNSR	The SDFFNSR is a negative-clock-edge triggered, static D-type flip-flop with scan input (SI), active-high scan enable (SE), and asynchronous active-low reset (RN) and set (SN). Set (SN) dominates reset (RN).
SDFFR	The SDFFR is a positive-clock-edge triggered, static D-type flip-flop with scan input (SI), active-high scan enable (SE), and asynchronous active-low reset (RN).
SDFFRHQ	The SDFFRHQ is a positive-clock-edge triggered, static D-type flip-flop with scan input (SI), active-high scan enable (SE), and asynchronous active-low reset (RN). The cell has a single output (Q) and fast clock-to-out path.
SDFFS	The SDFFS is a positive-clock-edge triggered, static D-type flip-flop with scan input (SI), active-high scan enable (SE), and asynchronous active-low set (SN).
SDFFSHQ	The SDFFSHQ is a positive-clock-edge triggered, static D-type flip-flop with scan input (SI), active-high scan enable (SE), and asynchronous active-low set (SN). The cell has a single output (Q) and fast clock-to-out path.
SDFFSR	The SDFFSR is a positive-clock-edge triggered, static D-type flip-flop with scan input (SI), active-high scan enable (SE), and asynchronous active-low reset (RN) and set (SN). Set (SN) dominates reset (RN).
SDFFSRH Q	The SDFFSRHQ is a high-speed, positive-clock-edge triggered, static D-type flip-flop with scan input (SI), active-high scan enable (SE), and asynchronous active-low reset (RN) and set (SN). Set (SN) dominates reset (RN). The cell has a single output (Q) and fast clock-to-out path.
SDFFTR	The SDFFTR is a high-speed, positive-clock-edge triggered, static D-type flip-flop with scan input (SI), active-high scan enable (SE), and synchronous active-low reset (RN). Scan enable (SE) dominates reset (RN).
SEDFF	The SEDFF is a positive-clock-edge triggered, static D-type flip-flop with scan input (SI), active-high scan enable (SE), and synchronous active-high enable (E).
SEDFFHQ	The SEDFFHQ is a positive-clock-edge triggered, static D-type flip-flop with scan input (SI), active-high scan enable (SE), and synchronous active-high enable (E). The cell has a single output (Q) and fast clock-to-out path.
SEDFFTR	The SEDFFTR is a positive-clock-edge triggered, static D-type flip-flop with scan input (SI), active-high scan enable (SE), synchronous active-high enable (E) and synchronous active low reset (RN). Scan enable (SE) dominates reset (RN) and enable (E).
TLAT	The TLAT is an active-high D-type transparent latch. When the enable (G) is high, data is transferred to the outputs (Q, QN).
TLATN	The TLATN is an active-low D-type transparent latch. When the enable (GN) is low, data is transferred to the outputs (Q, QN).
TLATNR	The TLATNR is an active-low D-type transparent latch with asynchronous active-low reset (RN). When the enable (GN) is low, data is transferred to the outputs (Q, QN).
TLATNS	The TLATNS is an active-low D-type transparent latch with asynchronous active-low set (SN). When the enable (GN) is low, data

	is transferred to the outputs (Q, QN).
TLATNSR	The TLATNSR is an active-low D-type transparent latch with asynchronous active-low set (SN) and reset (RN), and set dominating reset. When the enable (GN) is low, data is transferred to the outputs (Q, QN).
TLATR	The TLATR is an active-high D-type transparent latch with asynchronous active-low reset (RN). When the enable (G) is high, cdata is transferred to the outputs (Q, QN).
TLATS	The TLATS is an active-high D-type transparent latch with asynchronous active-low set (SN). When the enable (G) is high, data is transferred to the outputs (Q, QN).
TLATSR	The TLATSR is an active-high D-type transparent latch with asynchronous active-low set (SN) and reset (RN), and set dominating reset. When the enable (G) is high, data is transferred to the outputs (Q, QN).
TTLAT	The TTLAT is an active-high D-type transparent latch with active-high output enable (OE). When the enable (G) is high and the output enable (OE) is high, data is transferred to the output (Q).

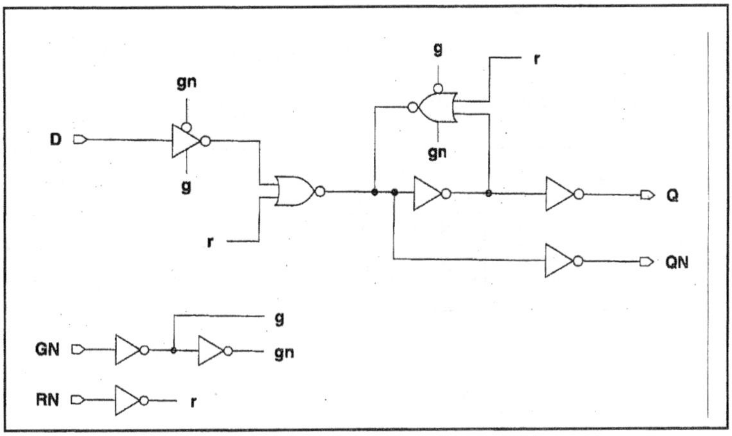

Figure 3-2. Active-Low D-Type Transparent Latch with Asynchronous Active-Low Reset [59].

Table 3-2. Truth Table for Latch Shown in Figure 3-2 [59].

RN	GN	D	Q[n+1]	QN[n+1]
1	0	0	0	1
1	0	1	1	0
1	1	X	Q[n]	QN[n]
0	X (Don't care: 0 or 1)	X	0	1

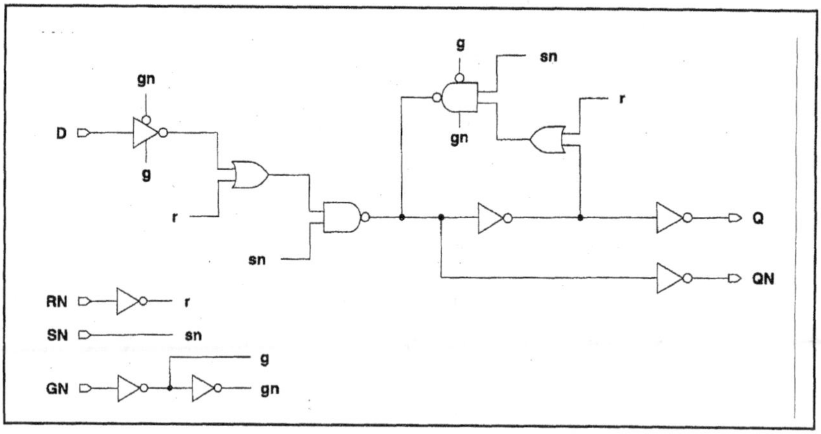

Figure 3-3. Active-Low D-Type Transparent Latch with Asynchronous Active-Low Set and Reset [59].

Table 3-3. Truth Table for Latch Shown in Figure 3-3 [59].

RN	SN	GN	D	Q[n+1]	QN[n+1]
1	1	0	0	0	1
1	1	0	1	1	0
1	1	1	X	Q[n]	QN[n]
0	1	X	X	0	1
1	0	X	X	1	0
0	0	X	X	1	0

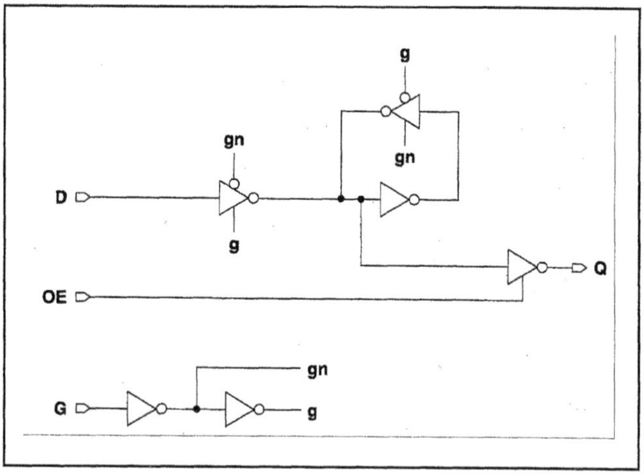

Figure 3-4. Active-High D-Type Transparent Latch with Active-High Enable [59].

Table 3-4. Truth Table for Latch Shown in Figure 3-4 [59].

OE	G	D	Q[n+1]
0	X	X	Z (high impedance)
1	1	0	0
1	1	1	1
1	0	X	Q[n]

Figure 3-5 shows a RS-type latch with active-high set (S) and reset (R), and Table 3-5 shows the truth table. Instead of the feedback loop in the previous latch structures, the RS-type latch uses the coupled NAND gates to implement the latch function. When set (S) and reset (R) are at different binary values, the set (S) value is passed to the output (Q[n+1]) of the latch. When using NOR gates instead of NAND gates in the RS-type latch structure, as shown in Figure 3-6 and Table 3-6, the reset (R) value is passed to the output (Q[n+1]) of the latch. This kind of RS-type latches does not require a clock signal.

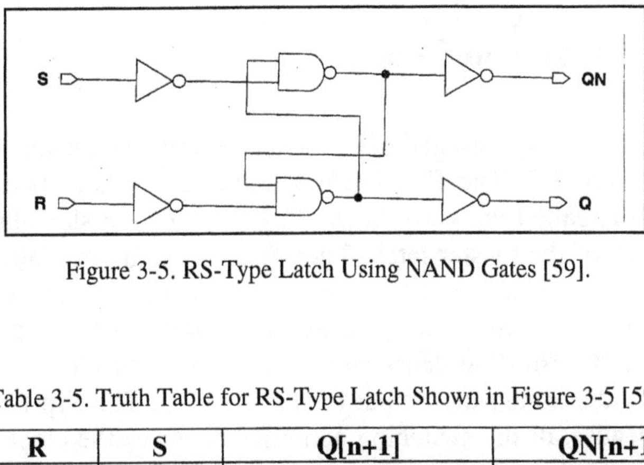

Figure 3-5. RS-Type Latch Using NAND Gates [59].

Table 3-5. Truth Table for RS-Type Latch Shown in Figure 3-5 [59].

R	S	Q[n+1]	QN[n+1]
0	0	Q[n]	QN[n]
0	1	1	0
1	0	0	1
1	1	IL (Illegal Logic)	IL

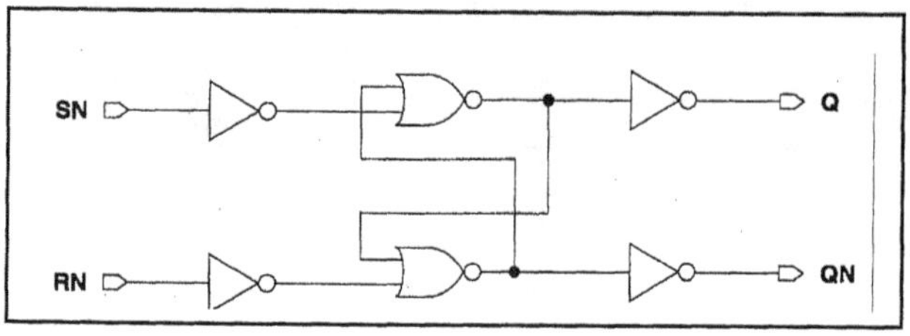

Figure 3-6. RS-Type Latch Using NOR Gates [59].

Table 3-6. Truth Table for RS-Type Latch Shown in Figure 3-6 [59].

RN	SN	Q[n+1]	QN[n+1]
0	0	IL	IL
0	1	0	1
1	0	1	0
1	1	Q[n]	QN[n]

3.2 Flip-Flop Clocking

A flip-flop is consisted of two latches connected back to back, as shown in Figure 3-7. The first latch is called the master latch and the second latch is called the slave latch. The clock of the slave latch is the inverted clock of the master latch. Therefore, if the master latch is open, the slave latch is closed, and vice versa. The output of the master latch passes through the slave latch to switch the output of the flip-flop. The input data of the flip-flop can change after the clock closes the master latch without disturbing the output of the flip-flop. The flip-flop is edge triggered because all the switching activities occur as the clock rises and the data will not change until the clock rises again.

Metastability can occur when the input data violates the setup and hold time of the latch or the flip-flop. Metastability describes the state of the latch or flip-flop with an undetermined output, which can be 0, 1, or X value hold. In order to resolve the metastability, an extra synchronizing or more settling time is allowed. Both approaches cost the time. One may not be able to afford the extra time and the best approach is not to violate the setup and hold time of the latch or flip-flop. The min-delay is the minimum logic delay between latches or flip-flops to guarantee that the latch or flip-flop would always capture the correct data. If the min-delay is violated, the data is propagated through the first latch and races through the second latch, while it is still open, thus corrupting the second latch output. Similarly, if the min-delay is violated for a flip-flop, the following flip-flop will capture the wrong data of the preceding flip-flop. The total logic delay between the latches or flip-flops has to be larger than the total clock skew and jitter between clocks driving these latches or flip-flops to avoid the min-delay violations.

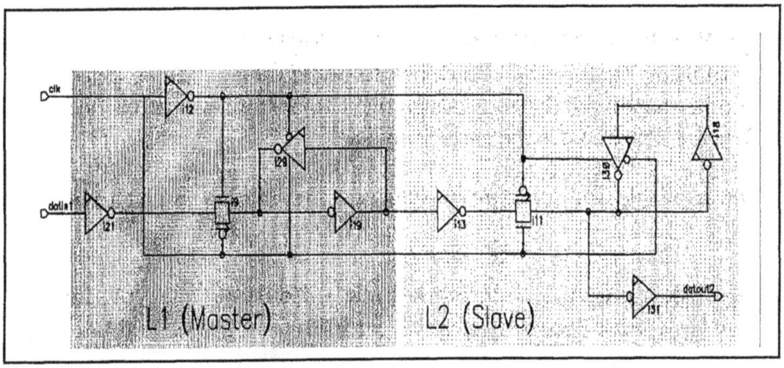

Figure 3-7. Flip-Flop Based on Master and Slave Latches [65].

Figure 3-8 shows the positive-edge triggered, static D-type flip-flop with scan input (SI) and active-high scan enable (SE). The truth table is shown in Table 3-7. When the SE signal is low (0), the flip-flop delivers the input data D at the rising clock edge. When the SE signal is high (1), the flip-flop is in the scan mode and the flip-flop delivers the scan input data SI at the rising edge. Two NAND gates and one inverter, as shown in Figure 3-8, are used to select the SI and D for SE high or low to be passed to the output Q and QN. The circuit shows an example of using scan input data for flip-flops. The chain of the scan inputs to one flip-flop, and then to the next flip-flop, and so on in the data path of the chip, is called a scan chain. The scan flip-flop is used to enhance the testability in the design. It feeds the scan input data and gets a second chain of data outputs from the data path. Without the inverter to QN, the output to Q will be faster because of the reduced load in the feedback loop for the second latch. Figure 3-9 shows a negative-edge triggered, static D-type flip-flop with scan input (SI), active-high scan enable (SE) and asynchronous active-low reset (RN). When the reset RN is high (1), the truth table as shown in Table 3-8. When the reset RN is low (0), the Q is set to 0 and QN to 1. Figure 3-10 shows a negative-edge triggered, static D-type flip-flop with scan input (SI), active-high scan enable (SE), and asynchronous active-low reset (RN) and set (SN). When RN and SN both are high, SN dominates RN and the Q is low and QN is high. The truth table is shown in Table 3-9. This flip-flip also needs the timing constraints for SI, SE, set, and reset signals for high-speed applications. Figure 3-12 shows a positive-clock-edge triggered JK-type flip-flop with asynchronous active-low reset (RN) and set (SN). The set (SN) dominates over the reset (RN). Table 3-11 shows the truth function table.

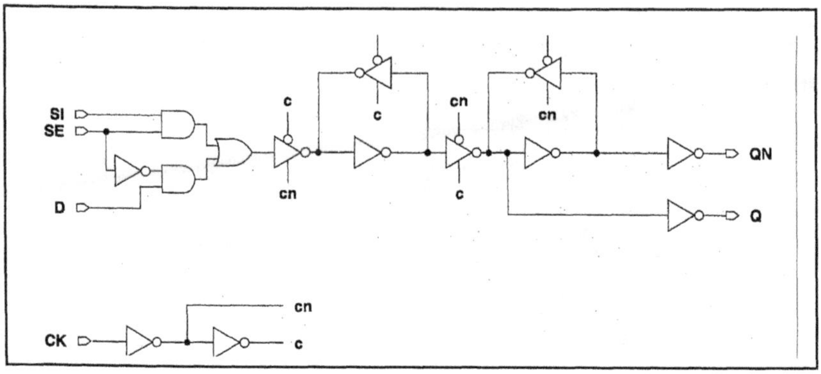

Figure 3-8. Static D-Type Flip-Flop with Scan Input [59].

Table 3-7. Truth Table for Flip-Flop Shown in Figure 3-8 [59].

D	SI	SE	CK	Q[n+1]	QN[n+1]
1	X	0	Rising	1	0
0	X	0	Rising	0	1
X	X	X	Falling	Q[n]	QN[n]
X	1	1	Rising	1	0
X	0	1	Rising	0	1

Table 3-8. Truth Table for Flip-Flop Shown in Figure 3-9 [59].

RN	D	SI	SE	CK	Q[n+1]	QN[n+1]
1	1	X	0	Rising	1	0
1	0	X	0	Rising	0	1
1	X	X	X	Falling	Q[n]	QN[n]
1	X	1	1	Rising	1	0
1	X	0	1	Rising	0	1
0	X	X	X	X	0	1

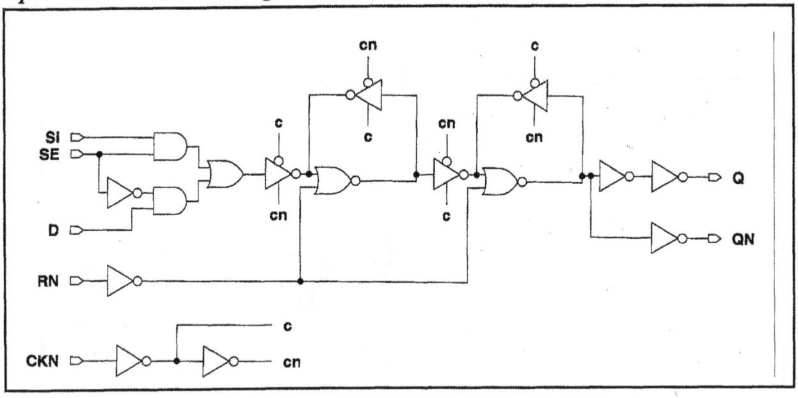

Figure 3-9. Static D-Type Flip-Flop with Scan Input and Reset [59].

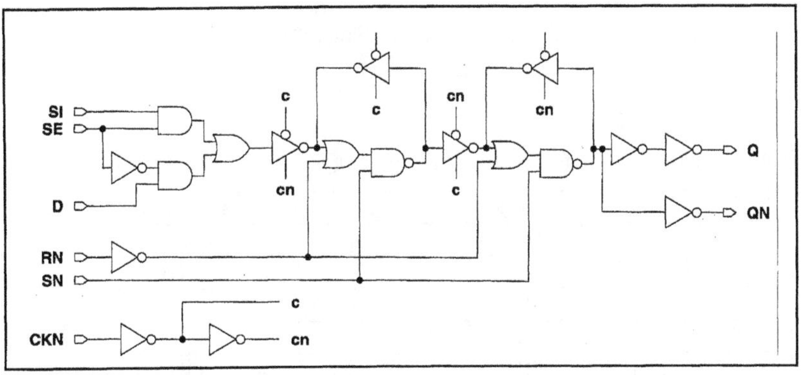

Figure 3-10. Static D-Type Flip-Flop with Scan Input, Reset and Dominant Set [59].

Table 3-9. Truth Table for Flip-Flop Shown in Figure 3-10 [59].

RN	SN	D	SI	SE	CKN	Q[n+1]	QN[n+1]
1	1	1	X	0	Falling	1	0
1	1	0	X	0	Falling	0	1
1	1	X	X	X	Rising	Q[n]	QN[n]
1	1	X	1	1	Falling	1	0
1	1	X	0	1	Falling	0	1
0	1	X	X	X	X	0	1
1	0	X	X	X	X	1	0
0	0	X	X	X	X	1	0

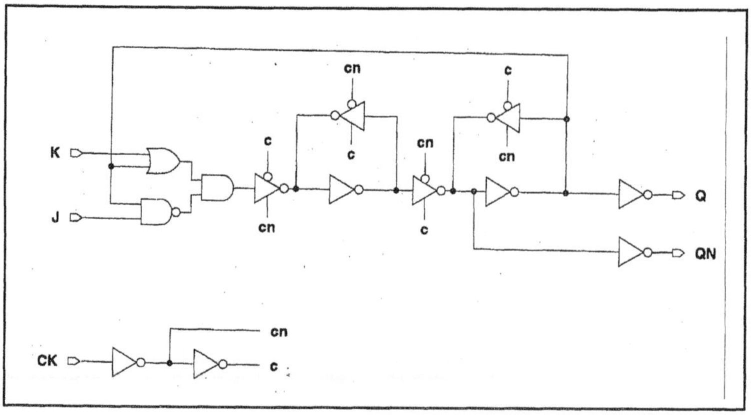

Figure 3-11. JK-Type Flip-Flop [59].

Table 3-10. Truth Table for JK-Type Flip-Flop Shown in Figure 3-11 [59].

J	K	CK	Q[n+1]	QN[n+1]
X	X	Falling	Q[n]	QN[n]
0	0	Rising	Q[n]	QN[n]
0	1	Rising	0	1
1	0	Rising	1	0
1	1	Rising	QN[n]	Q[n]

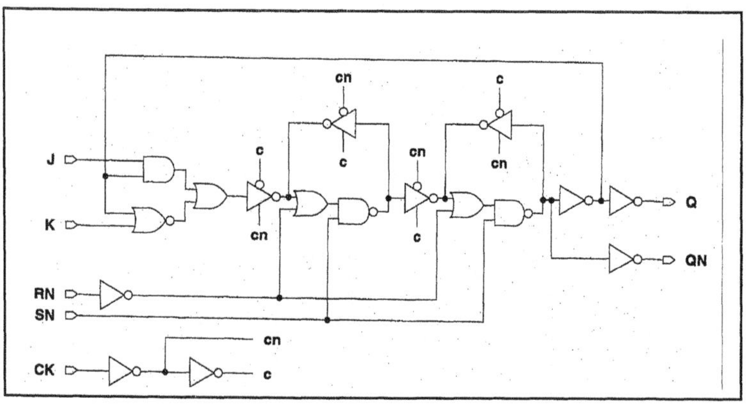

Figure 3-12. JK-Type Flip-Flop with Reset and Set [59].

Table 3-11. Truth Table for JK-Type Flip-Flop Shown in Figure 3-12 [59].

RN	SN	J	K	CK	Q[n+1]	QN[n+1]
1	1	X	X	Falling	Q[n]	QN[n]
1	0	X	X	X	1	0
0	1	X	X	X	0	1
0	0	X	X	X	1	0
1	1	0	0	Rising	Q[n]	QN[n]
1	1	0	1	Rising	0	1
1	1	1	0	Rising	1	0
1	1	1	1	Rising	QN[n]	Q[n]

3.3 Power Reduction

The clock network power consumption could consume up to a half of the entire microprocessor chip power based on the data shown in Figure 1-2. A large amount of the clock network power consumption goes to the clocked elements. In order to reduce the power in latches or flip-flops, a design exploration of the transistor sizes is needed. Figure 3-13 shows the trade-off between the device delay and device size. The key message is that for a device size $x > x_1$, the delay reduction is not significant but the power consumption increases because of the larger device sizes. We want to size down the "over-sized" devices in the circuit, while still satisfying the delay constraints, to reduce wasted power in the design.

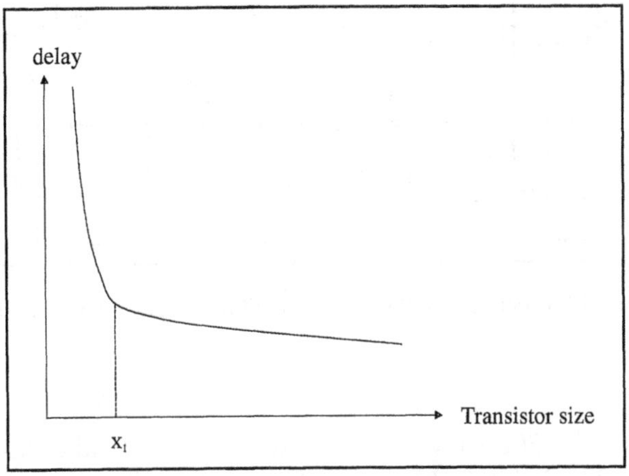

Figure 3-13. Delay Trend Versus Device Sizes.

To further reduce power consumption, a sequential logic circuit having a series of latches for datapath logic may be modified as follows [125]. Each data signal's latch is clocked using a corresponding qualified clock. The qualified clocks are generated by a series of AND gates that each gate has one input coupled to a global clock and the other input coupled to a valid bit, such that the latch is only clocked when a valid bit is present. Figure 3-14 shows an example using the proposed technique. The lower part shown in Figure 3-14 uses the traditional sequential circuits with the clk1, clk1#, clk2, and clk2# for the latches. The valid bit signifies that an instruction is being carried out in the sequential circuits. For example, in the instruction decoder of a microprocessor that sends the ADD, MULTIPLY, or DIVIDE instructions to the sequential circuit in the adder, multiplier, or divider, the upper part of the circuit contains a chain of latches, which generates the qualified clocks (clk1, clk1#, clk2, and clk2#) for the sequential circuit located in the lower part. The chain of latches in the upper part of the circuit continually transfers the valid bit in the pipe-stages. Inverters may be inserted in the chain of latches to avoid the min-delay problem due to the clock skew. The advantage of the above design is that the qualified clock is enabled only when the data operation is needed for a specific latch in the datapath. When the data is not delivered from the previous stages, the clocks are shut down. It is called the pipeline clock gating to reduce the power in the sequential logic.

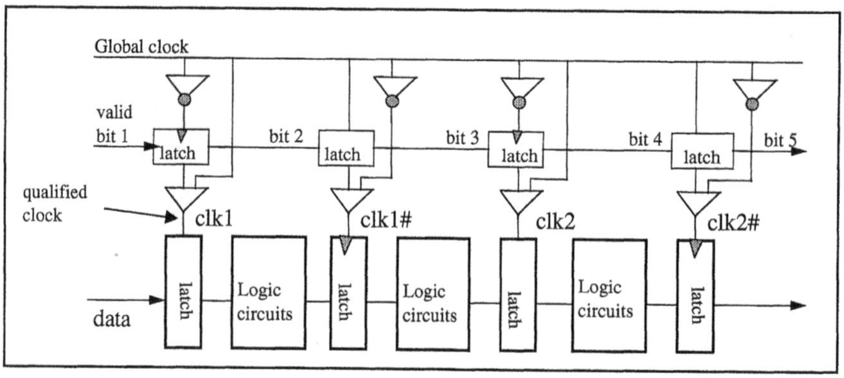

Figure 3-14. Clock Gating Based on Data Flow in Sequential Logic [125].

Figure 3-15(a) shows the logic waveforms of valid instruction bits (valid bit 1, bit 2, bit 3, bit 4 and bit 5). The qualified clock is high (enabled) only when the appropriate instruction valid bit is set to one. The qualified clocks (clk1, clk1#, clk2, and clk2#) are obtained based on the logical AND of clk or clk# with the relevant valid bit as follows: clk1 = clk & bit2, clk1# = clk# & bit 3, clk2 = clk & bit 4, clk2# = clk# & bit5. Figure 3-15(b) shows the logic waveforms of the qualified clocks. The qualified clock is high for a short period based on the instruction valid bit. The qualified clocks of latches are sequentially enabled based on which of the pipe-stages is being processed in the sequential circuit. The qualified clock of a latch becomes low when the data has been idle at the current pipe-stage to prevent any wasted clocking power.

The last power reduction recommendation is to avoid using low threshold voltage (Vt) devices in the sequential logic if possible, since the leakage currents will increase significantly with low-Vt devices [76].

3.4 Summary

This chapter describes the basic circuits and the working principles for sequential elements including the latches and flip-flops. The latches are devices that are sensitive to the clock level. During the clock high or clock low phase, depending on the latch type, any input data change at the latch will cause the output data to change. The above working principle of the latch makes it possible for the time borrowing between latches. The flip-flop is two latches connected back-to-back with non-overlapped clock phases for each latch. The flip-flop is sensitive only to the rising or falling edge of the clock. Any input data changes for the flip-flop will be delivered to the output only at the clock rising or falling edges, depending on the type of the flip-flop. During the period of unchanged clock level (high or low), and input data change will not change the output of flip-flops. Flip-flops are not able to take advantage of any timing borrowing opportunity. This chapter explains various latches and flip-flops with clock inputs based on a standard cell library [59]. The latches are combined with the scan, reset, or set capabilities for more powerful functions and improving the testability in the design. The power consumption of the latch or flip-flop is a big concern in the high-speed design, because the clocking power takes a large portion of the entire chip power consumption. We describe two methods to reduce power consumption: (a) transistor sizing for the trade-off between the delay or speed and the transistor power or area; and (b) doing the clock gating based on the data valid flow.

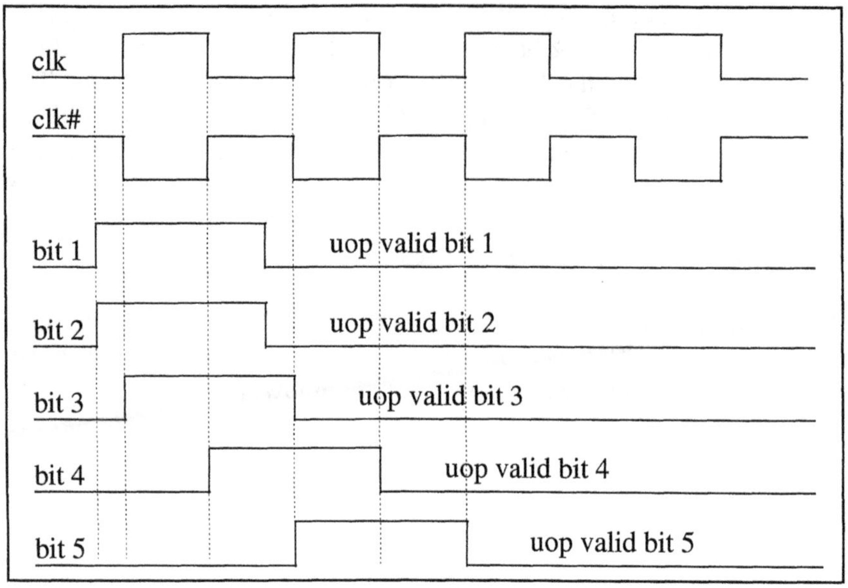

(a)

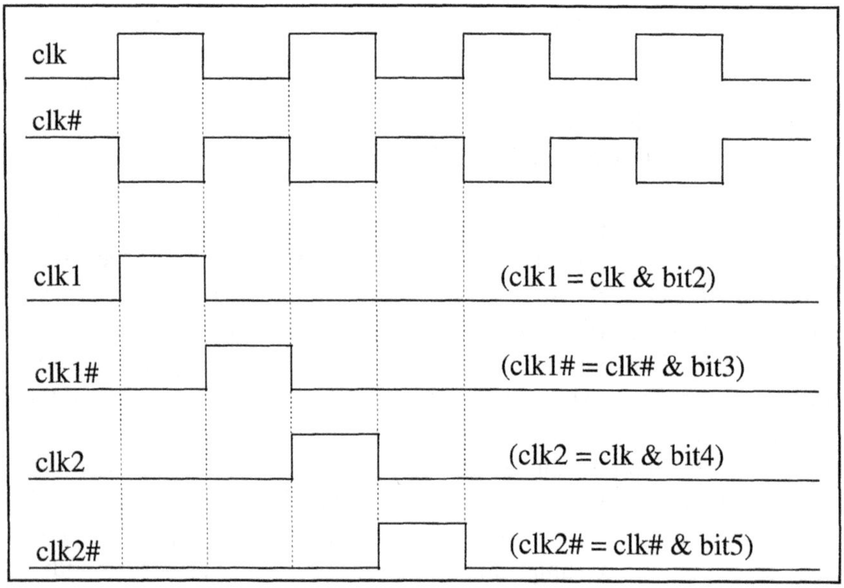

(b)

Figure 3-15. Logic Waveforms of Valid Data Bits and Qualified Clocks [125].

Chapter 4 Design Methodology for Domino Circuits

The domino circuit operates in two-clock phases. In the first phase, the clock pre-charges the output via a PMOS transistor to a high state (one). This is called the pre-charge phase. In the second phase, the NMOS logic decides if the output should be at a low state (zero) or kept at the high state (one). This is called the evaluation phase. Compared to the static CMOS circuit using dual NMOS and PMOS transistors to implement the logic, the domino circuit has the advantages of faster speed and smaller layout, because the domino uses fewer PMOS transistors. But the domino circuit needs special local clock distribution and consumes more power. Section 4.1 describes the domino operations. Section 4.2 discusses the clock distribution for domino circuits. Section 4.3 discusses further design optimization for domino circuits. Section 4.4 describes the low-Vt devices used in domino circuitry with considerations to the noise constraints. Section 4.5 contains a summary to this chapter.

4.1 Domino Circuit Types

Between the flip-flops and latches, the CMOS logic can be implemented in two ways. The most commonly used is the static CMOS circuit. The next alternative is the domino circuit, of which there are two types: (a) NMOS domino, and (b) PMOS domino. These are shown in Figure 4-1. The PMOS domino circuit is seldom used in circuit design. Most of the time, the NMOS domino is used because NMOS transistors are faster than PMOS for the same transistor widths. In the N-domino implementation style, for m inputs, the circuit needs 1 PMOS and m+1 NMOS transistors. In the static CMOS implementation style, for m inputs, the circuit needs m PMOS and m NMOS transistors. The N-domino circuit obviously needs fewer transistors compared to static CMOS implementation, when the inputs $m > 2$. This has a significant impact on the layout area and the capacitance loads for the earlier driving stage. There are several disadvantages for using the domino implementation, such as the power and noise issues, which will be discussed later. The domino circuit is riskier in the noise and clocking issues compared to the static CMOS.

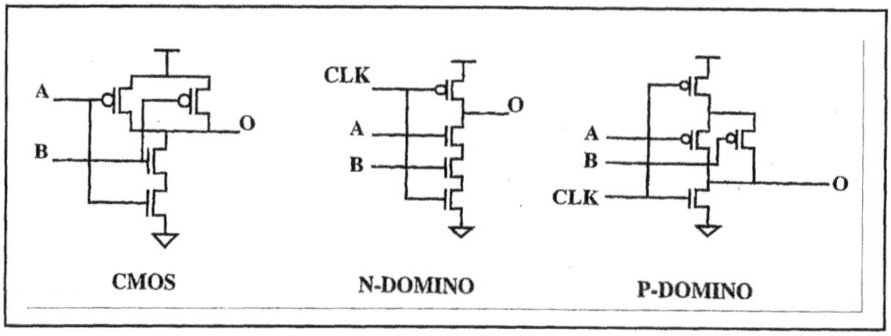

Figure 4-1. Static CMOS and Domino Circuits for NAND Logic.

Figure 4-2 shows two types of domino circuits: D1 and D2. D2 domino has the clk connected to the PMOS transistor. D1 domino has the clk connected to both PMOS and NMOS transistors for cutting off the pull-down path completely during the pre-charge phase. The extra n-device in the N-stack of D1 domino is controlled by clk to prevent the possibility of DC current paths during the pre-charge phase. The extra n-device increases the transistor sizes in the N-stack and also increases the clock loads. The advantage of D1 domino with the extra n-device connected by clk is that there are no restrictions on the input signals during the pre-charge phase. For the D2 domino, without the clocked n-type device, a DC path from Vcc to Vss is possible during the pre-charge phase. In order to eliminate the DC path to Vss, at least one input in the N-stack in every path from the output to Vss must be in the low state during the pre-charge phase. Removing the clocked n-device speeds up the domino circuit by about 10-20%. The D2 domino therefore has faster speed and uses a smaller area. The reduced sizes of the N-stack transistors help lower the input capacitance to the earlier driving circuit. The reduced clock loads also result in a faster transition and lower clock network power consumption.

For the D1-type domino circuit as shown in Figure 4-2, the domino circuit operates in two phases: the pre-charge phase and the evaluation phase. During the pre-charge phase when the clock is low, the output node O is pulled up by PMOS transistor i0 to Vcc (high state for O) by the low clk input. During the evaluation phase when the clock is high, the input signals can conditionally pull down the output to Vss (low state for O). The output O represents the logic implemented by the NMOS transistors in the evaluation phase. These NMOS transistors are called the N-stack in the domino circuit. Notice that in the pre-charge stage, the output is always pulled up even though the input signals try to pull the output down. One design guideline is to keep the domino output at the high logic state in most cases. This prevents constant toggling from the output high state to the low state, thus reducing the power.

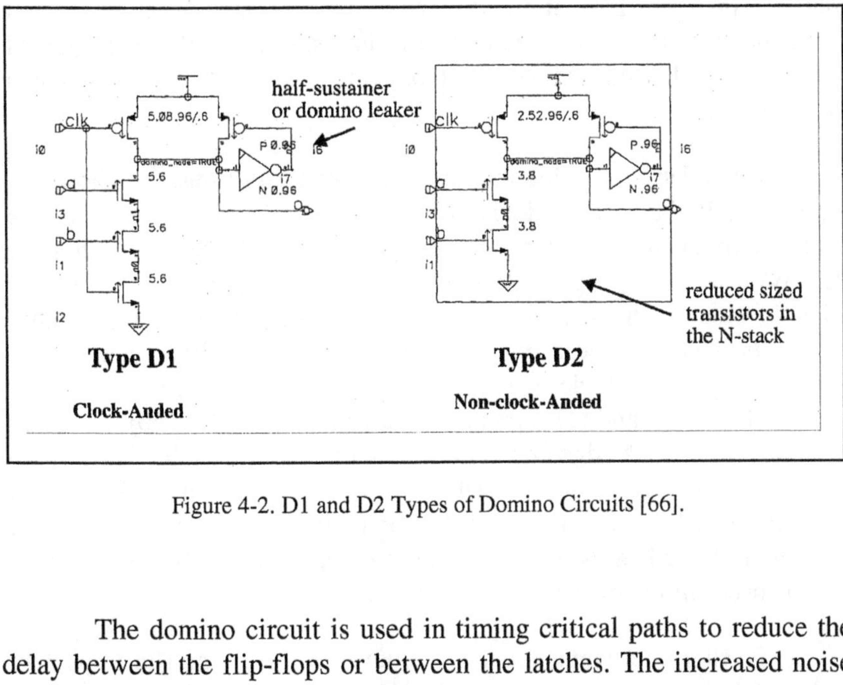

Figure 4-2. D1 and D2 Types of Domino Circuits [66].

The domino circuit is used in timing critical paths to reduce the delay between the flip-flops or between the latches. The increased noise and clocking problems make it difficult to design. But the domino circuit is still preferred in the tight timing window of the data path circuit. Here is an example how a critical signal may ripple through a domino path. During the evaluation phase, the high input signal of a domino circuit causes the output node to pull down to Vss. The output signal of this stage domino is inverted to high by an inverter and the high signal will cause the next stage domino circuit to pull down to Vss. The same sequence will occur for additional domino stages, and the signal will ripple through like dominoes.

There are some design concerns for the transistor sizes. The PMOS transistors, with their smaller sizes, are preferred to those in N-stack transistors. This is called the low-skewed domino size design. The size skew is defined as the size of an NMOS transistor to that of a PMOS transistor. The smaller PMOS transistor results in faster pull-down transitions and also smaller load capacitance. The smaller load capacitance improves the speed of the data path. The inverting gates between the domino stages have sizes optimised for smaller delays. The gates are sized to make faster low-to-high transitions. Thus, the inverting gates are hi-skewed to pull up faster with the PMOS transistors sized bigger than the NMOS transistor, in order to achieve fast low-to-high transition. The hi-skewed gates have high trip points. When the pre-charged output node begins to discharge, the trip point is detected immediately at the input to the hi-skewed inverter that makes a fast low-to-high transition. This allows the next stage domino to evaluate and ripple the data faster. In summary, the low-skewed domino is coupled with the high-skewed CMOS inverters or high-skewed CMOS NOR and NAND gates.

Consider a D1 domino used between flip-flops, as shown in Figure 4-3. If the data is clocked on the rising (or falling) edge of the clock, the inputs to the domino will not be ready until after a certain delay (Tdelay) after the rising (or falling) edge of the clock. Thus, one must delay the clock to the domino gate so that its evaluation phase begins after the inputs have become stable. This leaves less than 50% of the clock cycle to do the real logic evaluation. The improvement to the clocking between flip-flops and D1 domino is shown in Figure 4-4 [66]. We latch the data at one clock edge and begin evaluation of the domino logic at the opposite clock edge to ensure that the flip-flop outputs are stable. Static combinational logic functions can be completed before the evaluation of the domino logic begins. In this clocking scheme, about 50% of the clock cycle is used for the evaluation of the logic.

Further improvement can be achieved by replacing flip-flops with latches. Latches are transparent or level-triggered such that the data can flow through the latch when it is open and hold the data when it is closed, as shown in Figure 4-4. During CLK low, latch 1 is open while latch 2 is closed. The stage-2 domino chain is evaluating while the stage-1 domino chain is pre-charging. Latch 1 is open and allows the data to flow through the latch. As long as the domino stage is in the pre-charge state, the D1 inputs from the latch will be irrelevant. When the CLK is changed from low to high, the data is immediately ready from latch 1 as the inputs to the first domino stage. This is an advantage over the master-slave flip-flop that has to wait for the data to be stable. Latch 1 is closed and the domino stage-1 begins evaluation, allowing the data to evaluate through the first domino chain. During this time, the stage-2 domino chain is pre-charging; and latch 2 is open, ready to capture the output of the stage-1 domino chain. The D2 type domino can also be used in a D1-D2-D2 sequence in each stage to optimise the path. But extra clocking requirements will be needed for D1-D2-D2 types of domino chain.

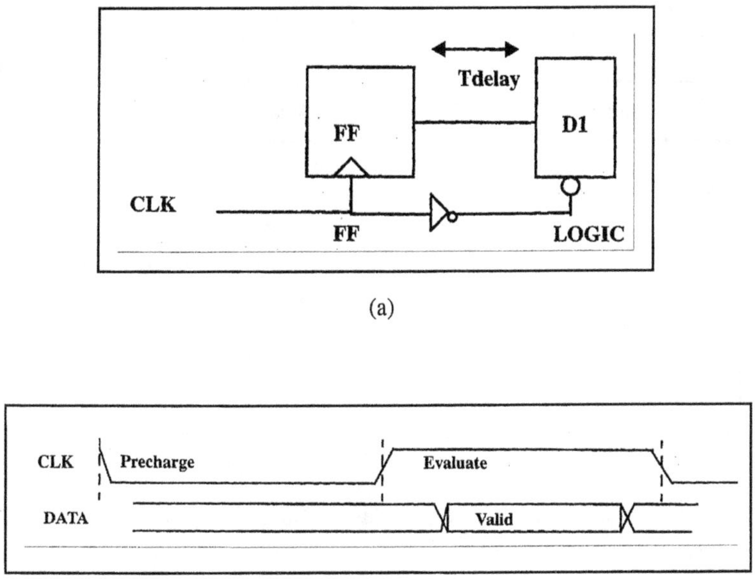

(b)

Figure 4-3. Flip-Flop Driving a D1 Type Domino [66].

In a D1-D2-D2 chain, if many D2 domino gates are connected in series, pre-charge time will run out since the clock must be delayed for each stage. By placing a D1 domino between a long series of D2 gate, we can make more use of the clock to complete longer domino chains. An example is shown in Figure 4-5. Both D1 gates A and D will pre-charge on the same clock edge CLK, and both domino gates B and E will be pre-charged on the delayed clock DCLK1, and finally both D2 domino gates C and F will be pre-charged on the even more delayed clock DCLK2. Ideally, the evaluation clock edge to each domino gate should begin at the same time or a slight delay with the data rippling through the data stages.

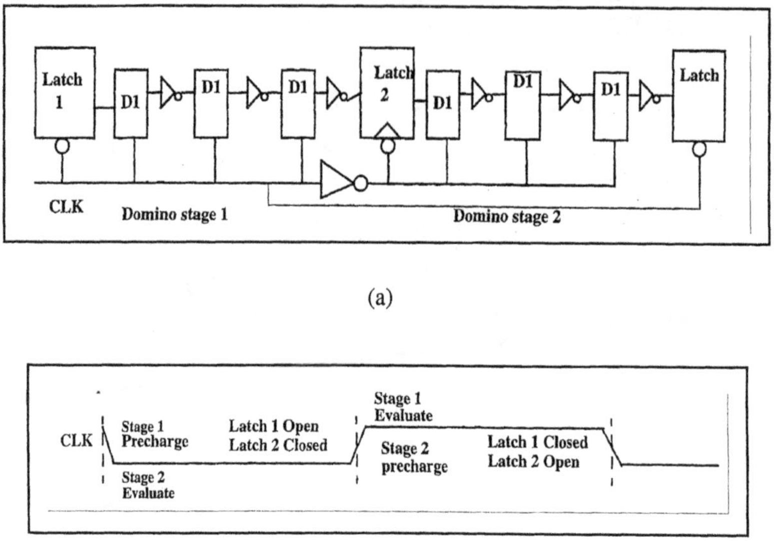

Figure 4-4. Latch Driving D1 Domino [66].

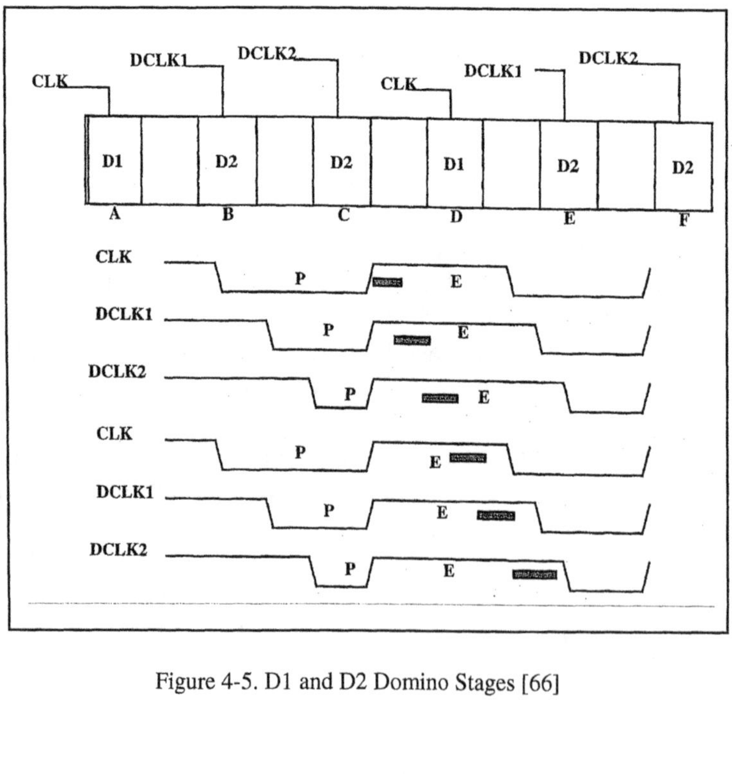

Figure 4-5. D1 and D2 Domino Stages [66]

4.2 Clock Distribution for Domino Circuits

A violation occurs in the domino design when the latch closes as the domino in the previous stage begins to pre-charge. In this case, the pre-charge clock edge to the domino may arrive too early and destroy the output data and violate the hold time of the latch. This would occur if the time to pre-charge the domino and the delay through the CMOS inverter is very short or when the domino directly drives the latch. In this condition, the pre-charge value (zero) will be latches instead of the correct value, as shown in Figure 4-6(a). To fix the min-delay problem, the pre-charge clock edge CLK to the domino is delayed to DCLK, as shown in Figure 4-6(b). Therefore, the evaluation phase is delayed to pass the data to the next stage latch in order to meet the hold time of the latch.

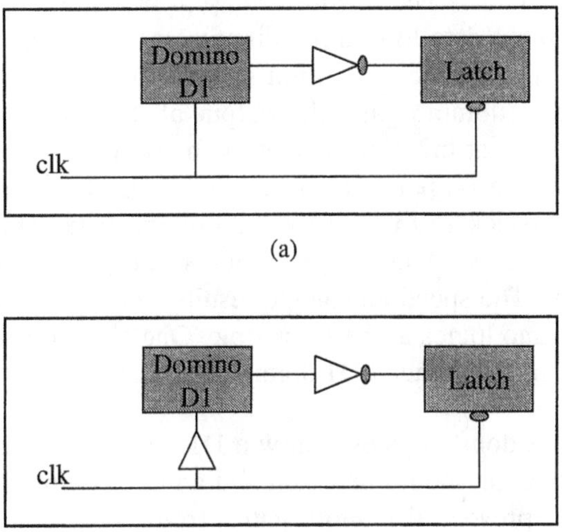

(b)

Figure 4-6. Fixing Min-Delay Violations.

It is preferable to use latches at the beginning and at the end of domino circuits. Compared to flip-flops, latches reduce the amount of dead evaluation time. A transparent latch driving the input of a domino chain must be closed during the evaluation phase. A transparent latch that captures the output of a domino chain should be open during the evaluation phase. Transparent latches should drive a D1 domino rather than a D2 domino in order to eliminate a possible DC path. The reason is that during the pre-charge of domino, the driving latch may still be open.

There is no guarantee that the latch's output will always be a 0 value (low state) that is required by D2 domino gates to prevent a DC path to Vss.

The D2 domino is usually followed with a high-skew inversion gate. Then, the inputs to the domino gate will be low during the pre-charging of the previous stage domino. The signal from the previous stage domino is pre-charged to high and inverted to low state by the CMOS inversion gate and the DC path is eliminated for the current stage D2 domino. The domino clock edges are often delayed to allow adequate setup time for the domino inputs. A late high-to-low transition will cause the output node of the domino to discharge. The inputs to the domino must be stable for a minimum setup time period before the evaluation phase begins.

Figure 4-7 shows the domino clocking methodology for three D1 type domino circuits. The evaluation and pre-charge phases of the DomCLK and input data to domino circuits are shown in Figure 4-9 for the waveforms. During the evaluation phase, with high DomCLK (phase 1) for the stage 1 domino gate, the output of this first stage domino is inverted from 0 to 1 at In2, thus activating the N-stack in the second stage domino gate, whose output is also inverted from 0 to 1 at In3, in turn activating the N-stack in the third stage domino gate. Therefore, in the evaluation phase of the DomCLK, the data is calculated in pipe-stages for one clock phase. The speed advantage results from doing the evaluations of the three domino logics at the same time. One clock line is used for the direct connections to cascaded D1 domino stages.

When D2 domino gates follow a D1 domino, as shown in Figure 4-8, extra precautions should be taken and a more complex clock scheme is necessary to prevent the short paths from Vcc to Vss in the D2 dominoes. Latches can only drive D1 dominoes. During pre-charge, the latch is still open and there is no guarantee that the latch output is a valid low required by the D2 domino. Only a previous domino followed by a high skewed inverter gate can drive a D2 domino, since the inputs to the second domino are guaranteed low during pre-charge. Successive D2 dominoes should begin pre-charging only after the previous D2 domino has finished the pre-charging, and the inputs to D2 have settled to a valid low to eliminate DC paths to Vss. That requires the pre-charge clock edge to arrive after the data inputs have fully pre-charged. This ensures that the N-stack completely turns off before the PMOS pull up transistor turns on, thus preventing the DC path from Vcc to Vss. Successive D2 gates should be in evaluation before the previous D2 gate has finished the evaluation, in order to eliminate a DC path to Vss. This requires that the evaluation clock edge must arrive before the data inputs have evaluated. Thus, the

PMOS pull up device can be turned off before the N-stack has a chance to turn on.

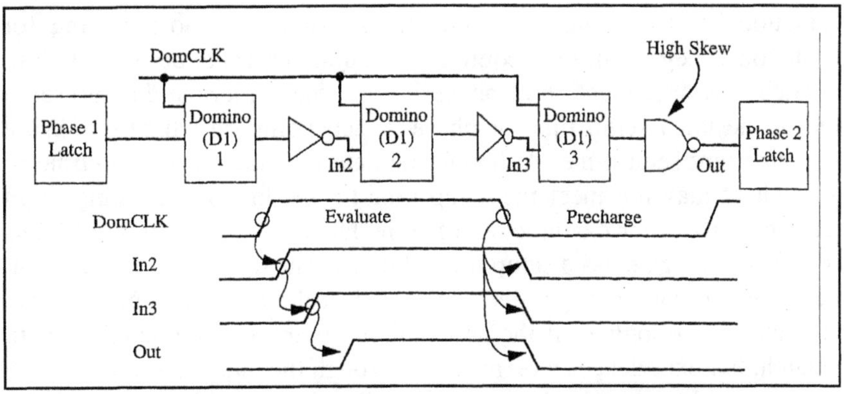

Figure 4-7. Clock Scheme in D1-D1-D1 Domino Chain [66].

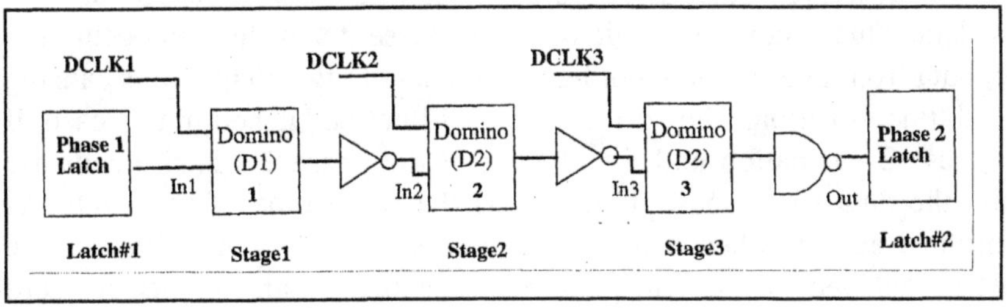

(a)

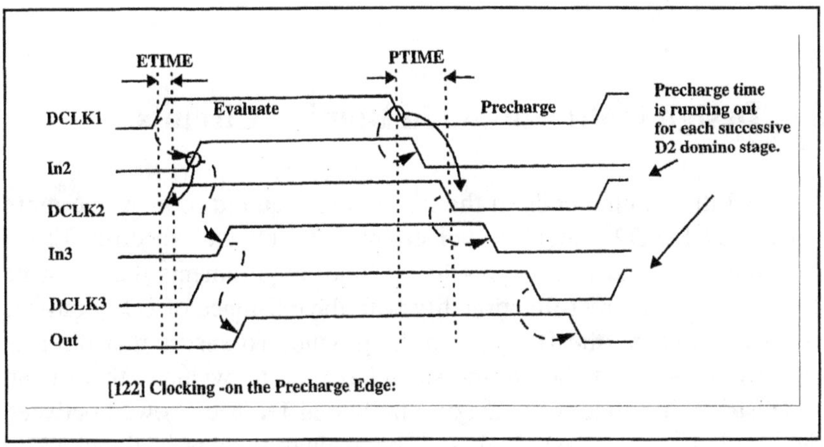

(b)

Figure 4-8. Clock Scheme in D1-D2-D2 Domino Chain [66].

We can eliminate the min-delay violations by holding the pre-charge edge until after the latch is closed. This allows the domino outputs to meet the hold time of the latch. The clock is designed so that the evaluation begins before the latch closes. No time is lost waiting for evaluation to begin, but this requires the domino inputs to be stable before the latch closes. Evaluation of phase 1 must meet either the domino setup time or the latch setup time of phase 2, depending on which one has a more stringent setup time. Although the data may meet the setup time of the latch, it may not meet the setup time to the domino, resulting in an incorrect output. For example, assume the latch has a setup time of 0.2ns and the domino gate has a setup time of 0.2ns. The data at node A as input to the domino gate is originally a logic 1. But the new data coming from the latch will be logic 0. If the new data meets the setup time of 0.2ns to the latch, but experiences a 0.1ns delay through the latch before it actually becomes stable at the latch output, it will only be stable at node A for 0.1ns and thus miss the setup time of the domino gate. The domino will evaluate incorrectly since it still sees its input as logic 1 rather than the new data, which should be logic 0. If the phase 2's evaluation occurs too late, data from phase 1 may not have been held long enough, thus causing a hold time violation. The correct value will not be latched in this case. If the phase 2 evaluation and the closing of the phase 2's latch occur too early, the data from phase 1 may miss the setup time of the latch. As mentioned earlier, when using a D2 domino, the clock pre-charge edge must be delayed in order to guarantee that the domino inputs are zero during the pre-charge. This prevents DC paths to the ground.

4.3 Design Optimization in Domino Circuits

All the domino cells in the library are required to meet the more stringent D1-D2-D2 clocking constraints. There are specific Etime (evaluation time) and Ptime (pre-charge time) requirements that must be met to ensure the correct functionality with the existence of D2 dominoes in order to eliminate the DC path to the ground. To ensure that the pre-charge always turns off before the stack has time to evaluate, there is an Etime restriction. Etime is the maximum time difference allowed between the evaluation edge of one clock and the evaluation edge of the clock in the next domino stage. The evaluation edge of subsequent domino stages must transition before the inputs evaluate high. This guarantees that there will be no overlap between the pre-charge phase of the clock and the discharge of the N-stack, thus preventing any DC paths to the ground. The pre-charge of the next domino stage cannot begin until the previous stage has been pre-charged and until the output of the CMOS inverting gate transitions low (the required input to the D2 domino pre-charge phase). Ptime is defined as the minimum amount of time needed between the pre-charge edge of DCLK1 and the pre-charge edge of DCLK2 of the next domino stage. This ensures that the N-stack will be turned off before the pre-charge of the cell begins, and thus prevents a DC path to the ground.

Many domino circuits are custom designed to drive the fanout. Following are the guidelines for custom domino design. All output nodes must have a keeper to prevent charge loss due to the leakage current at the output node. The worst-case charge sharing cannot exceed 15% of Vcc when the worst-case charge sharing occurs with fast inputs and the minimum output capacitance. At the domino outputs, the coupling noise cannot exceed 10% of Vcc. At the domino inputs, coupling noise should not exceed 25% of Vcc. More coupling noise is allowed at the domino input since it is driven by an active high-skew driver. If a noise source produces an upward spike, the equivalent resistance of the high-skew driver will help pull down the node, and prevent a noisy spike at the domino inputs. The smaller the RC, the better the input noise margins become. Nodes driven by large high-skew gates are able to cope with more coupling noise.

The coupling noise is the cross-talk effect that occurs when other wires are switching. If noise due to charge sharing exceeds 15% of Vcc, a secondary pre-charge device should be used. A secondary pre-charge device is needed if there is too much source-drain capacitance in the stack compared to the output. The secondary pre-charge device changes the intermediate node during the pre-charge phase, so that charge sharing during the evaluation phase is reduced. One example is shown in Figure 4-9. During the pre-charge phase, the output node is pre-charged high to Vcc. During the evaluation phase, input A transits from low to high as B remains at 0. Once A has changed from low to high, the charge stored originally in capacitor Cload is now re-distributed over Cload and CA. This will cause a drop in the output voltage that cannot be recovered. The waveform at the output node, shown in Figure 4-9, illustrates this drop. The smaller the Cload or the larger the CA, the more charge re-distribution and charge loss that occurs at the output node. The secondary pre-charge devices are used to help prevent charge sharing. They charge intermediate nodes during the pre-charge phase, so that during the evaluation phase, the worst-case output voltage dip due to the charge sharing is within a certain tolerance. The domino leaker, that is also called domino keeper, as shown in Figure 4-9, is a feedback loop from the output added to drive a weak PMOS pull-up device. The domino keeper is added to the output node in order to maintain the charge during the tri-state, when the output node must maintain its charge during the evaluation phase. It also helps recover the nodes after the noise has been injected due to charge sharing and coupling.

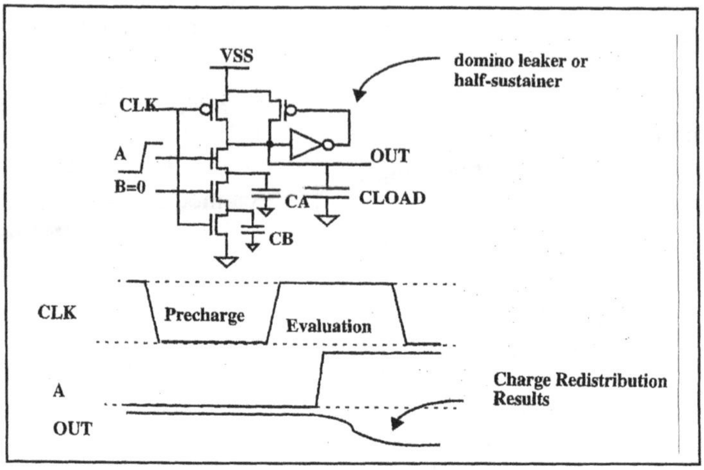

Figure 4-9. Charge Sharing in Domino Circuit [66].

4.4 Low-Vt Devices for Domino Circuits

The dual-Vt process adds the low-Vt (low voltage threshold) devices in the critical speed paths in order to reduce the delay or the total negative margin (TNM) [76]. The dual-Vt values in our experiments are as follows: NMOS with high Vt = 0.32V and low Vt = 0.24V, and PMOS with high Vt = 0.32V and low Vt = 0.24V in a 0.13μm process [76]. The Ids or delay improvement using the low-Vt device is expected to be about 15.5% for NMOS and 18.8% for PMOS. But the low-Vt devices significantly add leakage currents and noise problems, especially for the domino circuits. Reference [68] discusses the leakage current reduction with the dual-Vt devices. Reference [69] proposes a method for domino circuits to assign low-Vt devices on critical charging and discharging paths while the pre-charge transitions occur only through high Vt devices.

Figure 4-10 shows the low-Vt device candidates for a domino circuit. The candidates include NMOS-evaluation devices, CMOS driver, and CMOS receiver. The clock nodes, pre-charge nodes, and keepers in the domino circuit are not in the low-Vt candidates list, since they are not in timing critical paths. Note that the domino low-Vt candidates have low priorities compared to static circuit's low-Vt candidates, because the domino tends to increase the leakage current and noise more significantly. Inside a domino circuit, the elements are replaced in low-Vt based on the following order: CMOS driver -> domino NMOS evaluation devices -> CMOS receiver. The order is selected to update the output noise of each low-Vt replaced device. An output noise prediction model was developed for this purpose [76]. Within the required input noise thresholds, low-Vt candidates of domino circuits are selected for actual replacement in low-Vt devices. A noise threshold model was also developed based on various domino circuit structures and the number of parallel paths in the NMOS or PMOS stacks [76].

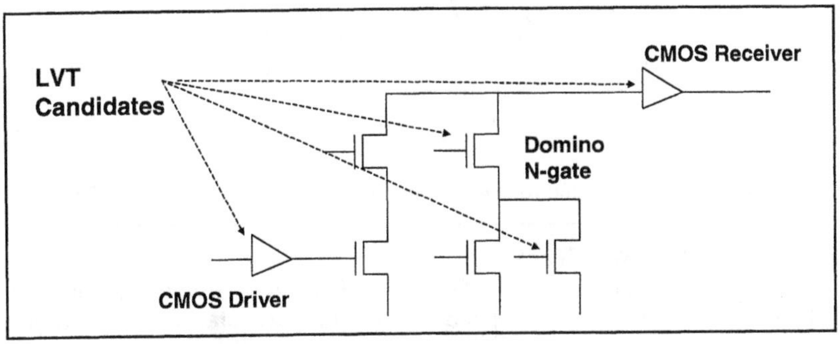

Figure 4-10. Low-Vt Candidates in Domino [76].

The NOR domino gate has the worse input noise impacts than the NAND domino gate, since the NOR domino has the parallel NMOS paths with the shorter paths from the output node to the ground. We made a set of noise measurements using the coupling noise simulation for a set of NOR domino gates, as shown in Table 4-1. The input noise to each NMOS transistor in the domino must be less than the noise threshold shown in Table 4-1. Otherwise, the excessive input noise will cause the output node to be pulled down to the wrong logic low. Table 4-2 shows the noise thresholds for NAND domino gates. One can see that the noise thresholds in NAND-type domino are much higher than the NOR-type domino. Table 4-3 shows the experimental results for microprocessor datapath blocks using the proposed low-Vt replacement models. The number of noise violations in the circuit is also shown in Table 4-3. The "static" column contains the results using the low-Vt devices for only static circuits in the timing critical paths; and the "domino" column shows the results using the low-Vt devices for all candidates related to domino circuits.

Table 4-1. Input Noise Thresholds for NOR Domino in 0.13µm Process [76].

NOR	2	3	4	5	6	7	8
Threshold (mV)	143	124	100	90	81	67	64

Table 4-2. Input Noise Thresholds for NAND Domino in 0.13µm Process [76].

NAND	2	3	4
Threshold (mV)	260	210	180

Table 4-3. Experimental Results of Microprocessor Datapath Blocks [76].

Block	**Max Neg. Slack static**	**domino**	**Total Neg. Margin static**	**domino**	**Noise Violations static**	**domino**
idfshrtn	-275	-275	-35407	-31030	72	74
idflongn	-302	-293	-47373	-45891	232	238
pfexcmp1d	-370	-370	-86006	-80582	4	14
pfexcmp0d	-396	-396	-149166	-143106	5	15
iemanctn	-323	-323	-92248	-91648	15	15
ieadjscd	-402	-402	-1744228	-1665518	33	46

An automated CAD flow, to replace high-Vt devices with low-Vt devices, is shown in Figure 4-11 [76]. It reads in the noise database and timing database, and then replaces low-Vt devices for the speed paths in an incremental approach. When a low-Vt device is replaced for a selected speed path, the delay and noise values are updated based on the polynomial models, as shown in Table 4-4. Then the next stage domino or receiver device will pick up the right noise with a low-Vt device in the earlier stage. In the noise update models as shown in Table 4-4, Y represents the delta of output noise from this low-Vt replaced device, and X is the input noise value. The models assume equal input noises for all NMOS stack transistors in the domino. The path number in Table 4-4 is defined as the possible input NMOS paths from the domino output node to the ground. For the example as shown in Figure 4-10, there are three possible paths from the output node to the ground through the NMOS transistors. The noise update models shown in Table 4-4 are constructed based on curve fitting to the set of input noise values with the average of low-Vt device replacement when other devices are already in low-Vt devices. They are worst-case noise update models since the low-Vt replacement flow really worry about the noise issues related to the domino circuits. The flow shown in Figure 4-11 is executed in two phases as follows:

(1) Without touching the domino and CMOS driver/receiver to the domino, perform the low-Vt replacement for the static circuits.

(2) Perform low-Vt replacements for the domino and CMOS drivers and receivers using the noise threshold models and noise update models.

In the static circuit low-Vt replacement (Phase 1), the noise is not a concern and only the speed path delay improvements are considered in the optimization goal. The noise models shown in Table 4-1, Table 4-2, and Table 4-4 are developed in a 0.13μm process, fast process corner, 1.45V voltage, and 110°C temperature. The fast process corner, high voltage, and high temperature lead to more noise in the circuit. In Phase 2, the initial delay or timing data for high-Vt circuit was obtained based on the PathMill static timing analysis tool from Synopsys [126]. The initial noise data was simulated based on a CAD tool from Intel, which includes the effects of the neighbouring lines couplings, the charge sharing, and the amplification of the noises through the logic gates [127]. The order of replacement for each domino stage is from the CMOS driver, to the domino N-stack, and then to the CMOS receiver. This way the noise in the next level will be updated incrementally for each low-Vt replacement. Thus the next level will pick the low-Vt device based on the updated noise value against the required thresholds.

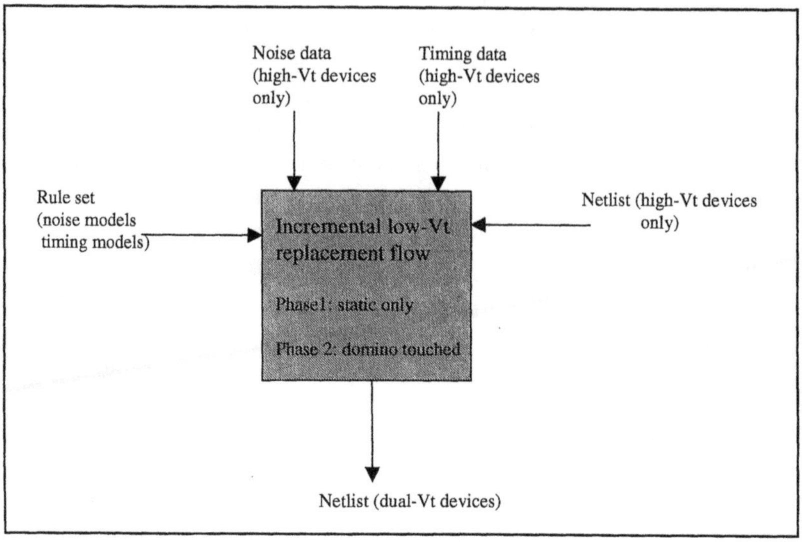

Figure 4-11. Low-Vt Device Replacement CAD Flow [76].

Table 4-4. Noise Update Models for Every Low-Vt Replacement [76].

X: Input noise, Y: Delta in output noise.

Path number	**Domino -> CMOS (NOR)**	**CMOS -> Domino (NAND)**
2	$Y = -0.0014X^2+1.0978X+0.0424$	$Y = 0.0003X^2-0.0475X+5.6802$
3	$Y = -0.0014X^2+1.0386X-05.9286$	$Y = 0.0002X^2+0.033X+2.249$
4	$Y = 0.0018X^2+0.5072X+8.8$	$Y = 0.00004X^2+0.1555X-4.5385$
5	$Y = -0.00005X^2+0.7307X+7.4583$	
6	$Y = -0.0025X^2+1.0567X+1.9758$	
7	$Y = -0.0017X^2+0.9297X+7.4424$	
8	$Y = -0.0023X^2+0.8956X+1.8727$	

4.5 Summary

The D1-D1-D1 domino chain has an easy clock scheme with all D1 domino gates using the same clock and is therefore easy for verification. The D1-D2-D2 domino chain has a speed improvement of 10% due to faster N-stack transition without a clocked N-gate in series in the D2 domino gates. But the clock scheme for the D1-D2-D2 domino chain is more difficult to design because one has to eliminate the possible DC path from Vcc to Vss when in the D2 domino is pre-charging. The N-stack needs to be low or completely off. Delayed clocks are required for the D2-D2 chain in order to meet the above pre-charge and evaluation time to eliminate the DC path to Vss. Domino logic has less silicon area compared to the static one, because fewer PMOS transistors are used in domino. A domino circuit usually consumes two to three times more power per function than a static circuit, because the domino output can switch every clock cycle. It is recommended that the domino circuit style be used when very high speeds are required to achieve zero speed paths such as in the microprocessor design. Noise is another concern in the use of dominoes, since the noise in the deep sub-micron increases because of the coupling noise in adjacent lines. Low-Vt devices must be used with care in the domino circuit. They will increase the leakage current and noise failures. We show one CAD flow in this chapter to deal with the low-Vt replacement in the domino circuit.

Chapter 5 Clock Generation and De-skewing

A phase-locked loop (PLL) or a delay-line loop (DLL) is a programmable, frequency synthesizer macro cell designed specifically for clock generation or de-skewing. A PLL or DLL can be used to adjust dynamically the edges of the output clock to match those of the input clock. PLLs can be used to double or multiply the input clock frequency, such as outputting the clock with 2X and 4X clock frequencies. DLLs, which are also called de-skewing buffers, are used to compensate the clock skew resulting from the delay mismatch in the clock network. The difference between the PLL and the DLL is that the DLL uses only the inverter delays, while the PLL contains the analog VCO (voltage controlled oscillator) for the clock phase adjustment. Section 5.1 introduces an on-chip clock generator. Section 5.2 presents the PLL characterization results. Section 5.3 provides the PLL floorplan guidelines. Section 5.4 describes DLL circuits and de-skewing buffers. Section 5.5 shows an on-die clock shrinking technique for silicon debug. The detailed circuits of the PLL interior components (charge pump, VCO, delay matching, divider, etc.) are not included in this book. They can be found in a good reference for PowerPC microprocessor [89].

5.1 On-Chip Clock Generation

The conventional phase-locked loop (PLL) consists of the following components: the phase detector, charge pump, loop filter, voltage-controlled oscillator, and divider circuit. The charge pump circuit and voltage-controlled oscillator are implemented using analog circuits for the current sources, etc. Therefore, the power supply should be dedicated to the analog circuit in order to reduce the supply noise and clock jitter. Figure 5-1 shows a block diagram of clock synthesizer or phase-locked loop in a 0.18μm process [79]. The synthesizer interfaces to a reference frequency and multiplies it up to the required clock frequency. The feedback path is separated to allow the clock buffer tree to be inserted. The PLL utilizes an internal loop filter and active bandwidth regulator to provide maximum noise immunity and minimum jitter. Table 5-1 lists the IO pins for the PLL shown in Figure 5-1. When in operation, the frequency of VFO is the same as that of the external reference clock, and the frequency of VFOX2 is twice that of the external reference clock. A detailed function table for the control pins PD, OL1, and OL2 are shown in Table 5-2.

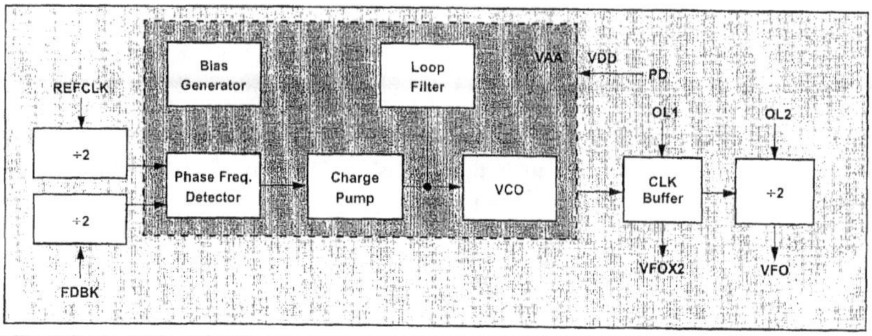

Figure 5-1. PLL Architecture [79].

Table 5-1. IO Pins of A Phase-Locked Loop [79].

Name	Characteristic/Direction	Description
REFCLK	Digital, Input	PLL reference frequency
OL1	Digital, Input	Output disable signal for VFOX2
OL2	Digital, Input	Output disable signal for VFO
PD	Digital, Input	PLL power down mode, logic 1: standby
VFO	Digital, Output	PLL div2 output clock
VFOX2	Digital, Output	PLL clock output
FDBK	Digital, Input	Feedback return of VFO after the clock distribution tree
Vdd	Power, Pad	Digital power
Gnd	Power, Pad	Digital ground
Vaa	Power, Pad	Analog power
Ag	Power, Pad	Analog ground

Table 5-2. Functions of A Phase-Locked Loop [79].

Pin Name	Type	Description
PD	PLL power down control	0: Normal operation. 1: VFOX2 = VFO = 0.
OL1	VFOX2 control	0: VFOX2 = normal operation. 1: VFOX2 = 0.
OL2	VFO control	0: VFO = normal operation. 1: VFO = 0.

The operation of a clock generator using the phase-locked loop is shown in Figure 5-2 [81]. When bypass asserts high and tcksel asserts high, the rest of the chip gets the scan clock. When tcksel asserts low, the rest of the chip gets the clock directly off the chip. When bypass asserts low, the reset of the chip normally gets the clock generated from the PLL macro except for a short transition period after the later resets (chiprst or pcirst) are de-asserted. During this transition, the rest of the chip gets the clock with half the frequency of padck. Any clock switching should be treated as a whole chip reset event. This means all the storage elements need to be reset and 1ms lock time is required. The clock control unit controls the lock time, and handles the clock switching between the off-chip clock and the PLL macro. The data loaded by the off-chip clock before PLL locking will be valid after the clock switches to the PLL macro clock.

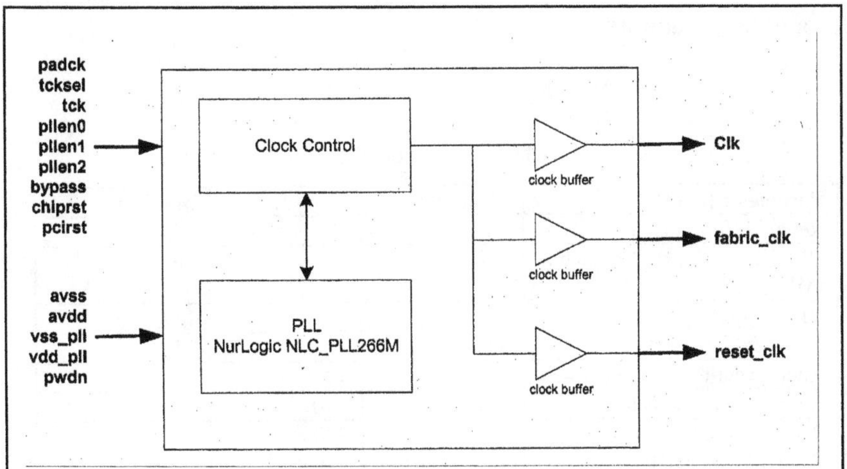

Figure 5-2. Clock Generator [81].

5.2 Characterization of Clock Generator

Notice that the PLL has been characterized for the clock jitter and acquire time. Jitter is defined as the clock edge shifting from the central position in multiple clock cycles; and the clock jitter of the PLL is the amount the source contributes to the clock instability for the entire on-chip clock distribution system. Therefore, minimizing the clock jitter due to the PLL is critical in the chip design. The power supply noise, especially that from the analog power supply for the PLL, contributes a lot to the clock jitter. In order to reduce jitter, minimize the power supply noise by adding a large amount of decoupling capacitors around the analog and digital power supplies to the PLL. The acquire time is defined as how long the phase-locked loop will be stable or locked to the output frequency. This is due to the time needed for the PLL circuit to change the output phase by getting the internal delays adjusted in order to align the output clock phase edge with the input or external clock phase edge. Shorter acquire time is useful for faster boost up in the clock to have the valid clock waveforms to the system. Table 5-3 shows the PLL analog AC characterization in a 0.18μm process for the circuit shown in Figure 5-1 [79]. Table 5-4 shows the supply voltage and noise thresholds (DC characterization) for this PLL [79]. In order to meet the indicated jitter specifications shown in Table 5-3, the 3.3V analog supply needs to be dedicated for this PLL, away from other power supplies. This analog supply can be generated from an on-chip voltage regulator or from the board 3.3V supply through a low pass filter with huge amounts of decoupling capacitors.

Table 5-3. AC Characteristics of NLC_PLL266M [79].

Parameter (Unit)	**Minimum**	**Typical**	**Maximum**
Reference/Input Clock (MHz)	50	-	133
VFOX2 Clock frequency (MHz)	100	-	266
VFO frequency (MHz)	50	-	133
Clock duty cycle (%)	45	-	55
Jitter (cycle-to-cycle, ps RMS)	-	10	-
Jitter (long term, ps RMS)	-	<40	-
Static phase error (ps), REFCLK -> FDBK	-	200	-
Acquire time (μs)	-	<20	100
Power (mW)	-	10	30

Table 5-4. DC Characteristics of NLC_PLL266M [79].

Parameter (Unit)	Minimum	Typical	Maximum
Analog Power (V)	3.0	3.3	3.6
Digital Power (V)	1.6	1.8	2.0
Analog Current (μA)	-	-	1
Digital Current (μA)	-	-	1

5.3 Layout Guidelines

The analog power supply is the first-order cause of the jitter. The analog power supply must be free of the voltage noise with frequency components near the loop bandwidth of the PLL. The reference clock frequency can be another external source of the jitter. The low frequency phase noise in the reference clock will pass through the PLL macro to the output, while the higher frequency reference noise will be attenuated. Given a good power supply and reference, it is important to have good layout, and good signal and supply routing to obtain optimal results. The most important layout factor is placement. The PLL should be placed in the quietest area of the substrate. Space should be provided between the PLL and any adjoining diffusion areas; 30mm is generally sufficient [82].

There is usually more noise in the center of the die, since the IR drop of the power supply adds to the inductive noise from the packaging. In addition, there is generally a higher level of substrate noise near high-speed output pads and high-speed un-terminated inputs. This is due to the I/O supply resonance coupling to the substrate, and to transmission line effects coupling through ESD protection. The use of isolating supplies from the I/Os and substrate does not completely eliminate these sources, but it can help to shunt some of the noise to external ground. Therefore, place the PLL near the pad ring and away from the fastest output buses. An area already isolated with analog supplies is optimal. The second choice would be a pad area made up primarily of input only pads. A corner made of input pads is worse than the middle of an edge consisting mainly of output pads.

Avoid placements directly next to large clock drivers or the drive sections of large buses. Memory arrays usually generate less noise than logic blocks. Placement along an array edge, away from the write or output circuits, is a good choice. Place the PLL close to the analog power pads that connect to the PLL's analog section. Keep the routing short and wide between the PLL and its analog power pads. Also, place the PLL near the core power and ground pads that connect to the PLL's digital section and the logic that interfaces to the PLL output and reference. Supply multiple PLLs from individual pads whenever possible. If not, use a separate route from the analog power and analog ground pads to each PLL.

Guard rings can provide immunity to substrate minority carriers, and can also help to reduce voltage gradients in the substrate. Most foundry design rules require a set of power and ground rings between the I/Os and the core logic. It is assumed that this type of guard ring is in place between the pad ring and any PLL. Nurlogic PLLs already contain a set of guard rings, which, when combined with all of the recommendations already given, are adequate to meet the expected jitter performance [79]. If a good PLL placement is not available, or if the choice is uncertain, then it is recommended that the white space be increased or additional guard rings inserted. Additional guard rings will be most effective if they are placed close to the circuitry adjoining the PLL. The composition of the above mentioned guard ring should involve a core power and core ground ring. Both rings should have fully contacted diffusion and metal. The outside ring should be N+ diffusion in an NWELL. It must be connected to the same core power as the surrounding circuitry and should be routed close to, and confirming with, that circuitry. Do not tie it to the PLL core power. Even a narrow NWELL ring is effective, and one 3mm wide is usually good enough. The ring towards the PLL should be a P+ diffusion substrate tie, connected to core ground, preferably the same ground feeding the digital portion PLL. Lastly, maximize the edge rates of the frequency reference and PLL output signals.

If multiple phase-locked loops are used for de-skewing the clock distribution buffers, the following checklist is suggested [80]. Make sure that a pair of dedicated digital power Vdd and Vss and analog power Vdd and Vss for each de-skew PLL exists. Do not share the digital power pins and analog power pins with other functional powers except the guard ring Vddg and Vssg. Make sure that Vddg and Vddg use the digital power pins during application. Since PLL performance is sensitive to the power noise, we need to make sure the power noise is within 10%Vdd for the digital power supply. We also minimize the ground noise for the analog power supply. During the placement and routing, make sure the feedback clock and the output clock from the PLL do not couple with other high-speed signals or go through other high-speed active area. Clock tree synthesis is preferred to minimize the path between the output clock and feedback clock to the PLL.

Figure 5-3 shows the distribution of the PLL. In Figure 5-3, the path delay from the PLL output driver to any function block should be the same for path A, B, and C. If there is a variation among different paths, one should keep the delay of the feedback path, such as path C, at a medium value with respect to those of other paths. The feedback clock paths should not be crossed or too close to other parallel lines. Make the bonding wires and package pins as short as possible to reduce the inductance effect at high frequency. If the chip is not pad-limited, use more power pads for the same power pin. Reduce the inductance effect at high frequencies by using the flip chip package. The output pads shown in Figure 5-3 are for testing. They can be turned off by the control logic in order to reduce the influence on the PLL normal operation.

5.4 De-Skewing Circuits

Figure 5-4 shows the de-skewing structure used in Intel IA-64 microprocessors, which is added at local clock buffers to reduce the clock skew [83]. The global clock distribution consists of two clock trees with the distribution of the core clock and the reference clock from the PLL clock generator to the de-skewing buffers. The second level in the clock distribution network is the regional clock distribution that consists of regional clock grids from the de-skewing buffers to 30 clock regions. The third level in the clock distribution network is the local distribution comprising the local clock buffers (LCBs) taking the input from the regional clock grid and the local interconnect to support the clocked elements and the opportunity time borrowing (OTB) delayed clock generation for domino circuits. A reference clock is distributed along with the core clock from the PLL clock generator to the de-skewing buffers (DSKs), as shown in Figure 5-4.

Figure 5-5 shows the de-skewing clock buffer structure and working mechanism consisting of the clock buffer and regional clock distribution grid. Each clock region is served with a dedicated de-skewing clock buffer. The local feedback clock is sampled from the regional clock grid and sent back to the de-skewing buffer. Other inputs to the de-skewing buffer are the global clock and the reference clock. A phase detector in the de-skewing buffer's local controller samples the phase difference between the reference clock and the local feedback clock. Adjustments to the core clock delay are accomplished through the variable delay circuit. The adjustment is continued until a minimum phase error is achieved. Any load mismatches and within-die variations in the core clock distribution are automatically compensated.

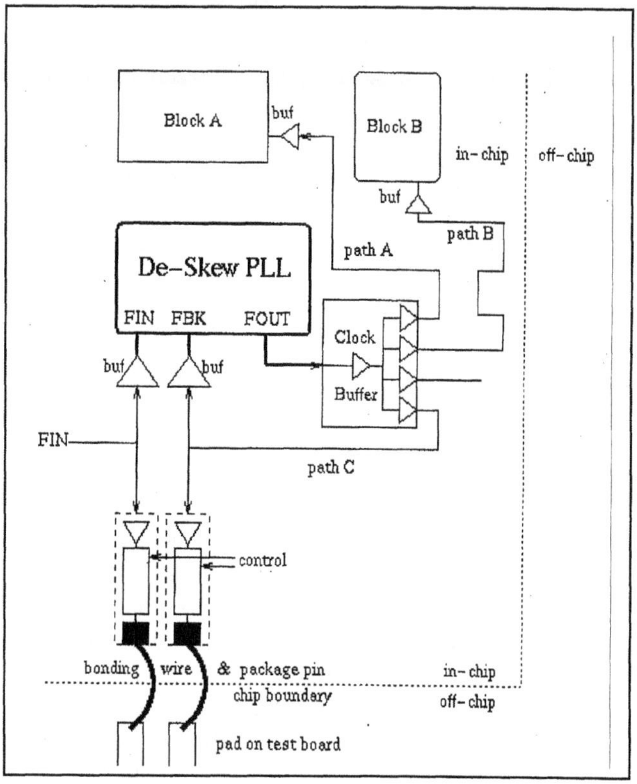

Figure 5-3. PLL Floorplan [80].

Another de-skewing buffer application can be found for the microprocessor clock distribution structure, as shown in Figure 5-6 [85]. The digital de-skewing circuit is used to equalize two clock spines by compensating for these mismatches and gradients. The circuit is composed of delay lines in both spines of the microprocessor distribution network, a phase detection circuit, and a controller. The phase detection circuit determines the phase relationship between the two spines and generates an output based on the phase relationship. The controller takes the phase detection information and makes a discrete adjustment to one of the delay lines that have been determined by the controller to require adjustment.

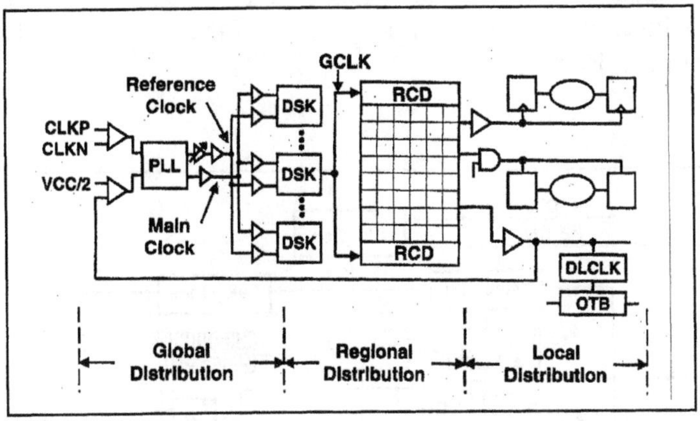

Figure 5-4. Microprocessor Clock Distribution Using De-Skewing Buffers (DSKs) [83].

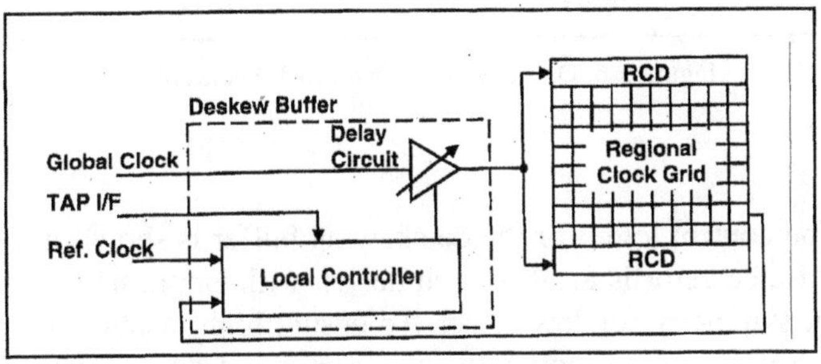

Figure 5-5. De-skewing Buffer [83].

Figure 5-7 shows the de-skewing buffer circuit structure with the variable delays [83]. It is a digitally controlled analog delay line consisting of a 20-bit delay control register, a two-stage variable delay circuit, and a push-pull style output buffer. The 20-bit delay control register forms a 20-step linear delay coding. A 20-bit delay control was selected based on the delay step-size resolution and the total buffer delay range. Delay adjustment can be accomplished by shifting a "1" from one end of the register to decrease its delay, or by shifting a "0" from the opposite end to increase its delay. The delay control register also accepts input from the test access port (TAP) interface. This feature permits a manual adjustment of the de-skewing buffer delay through the TAP interface. This interface can be used for post-silicon timing optimization. The capacitor loads are added alternatively to the two-inverter outputs. The use of two inverters allows the load to be split between two drivers and also provides a non-inverting delay. Both NMOS and PMOS loads are used in each of the capacitive loads in order to reduce the capacitive slope and balance the rising and falling edge of the inverters. The delay across the inverters will vary in accordance to the setting stored in the delay control register.

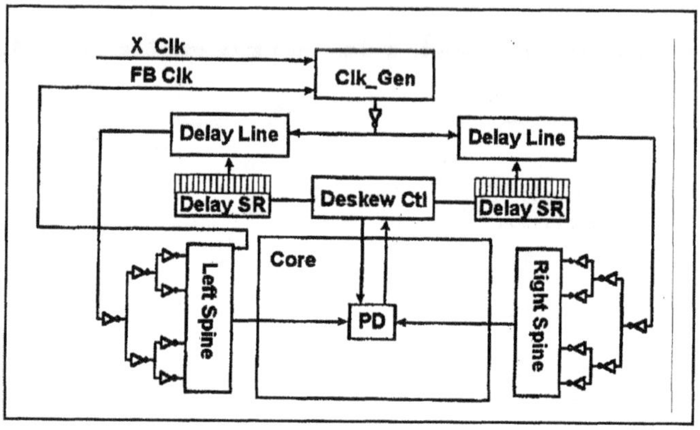

Figure 5-6. De-skewing of Two Clock Spines [85].

The control logic for the de-skewing buffer is shown in Figure 5-8. The phase detector is an R-S latch design with emphasis on the circuit and layout symmetry for low offset. To ensure high resolution and avoid meta-stability, the phase comparison is performed once every 16 cycles by an enable mechanism that samples the reference clock and feedback clock. By not having the phase comparison in every clock cycle, the phase detector is given plenty of time to resolve any phase differences. The output of the phase detector is a "Lead/Lag" signal that is further filtered by a six-tap digital low-pass filter before sending the result to the delay control register. The filter also detects a steady-state Lead/Lag pattern that signals the completion of the de-skewing operation. In this situation, the phase difference between the reference clock and the feedback clock will be less than one step size of the buffer delay. The local controller will stop the de-skewing operation and set the delay control register so that the feedback clock is always faster than the reference. This last procedure ensures that the overall residual skew due to the de-skewing buffers will be limited to one delay step size. The de-skewing operation starts during the initial microprocessor reset. The de-skewing for the clock regions proceeds in parallel. A global de-skewing controller monitors the progress and signals completion of the de-skewing. Once this occurs, the de-skewing buffer delay register settings will be fixed until the next power-up sequence [83]. Figure 5-9(a) shows the measured clock skews with and without the de-skewing mechanism in an Intel IA-64 microprocessor chip [83]. The total clock skew is 28ps with the de-skewing, and is four times larger without. The data is collected using the laser voltage probing (LVP) technique that enables high-resolution signal probing from the backside of the silicon die. Figure 5-9(b) shows the jitter measurement results with the function of switching events at the output of the clock buffering [83].

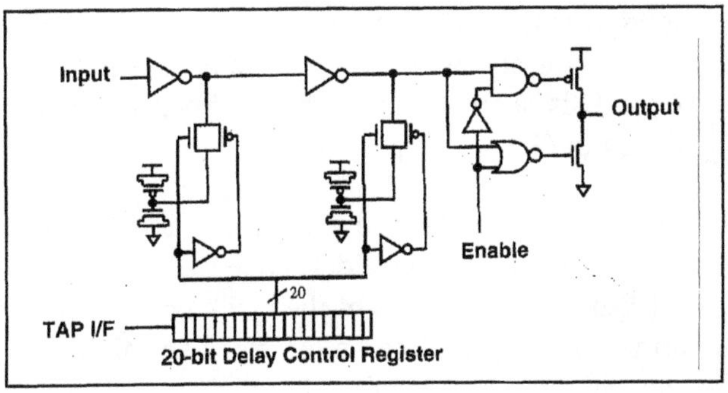

Figure 5-7. De-skewing Buffer With Variable Delay Lines [83].

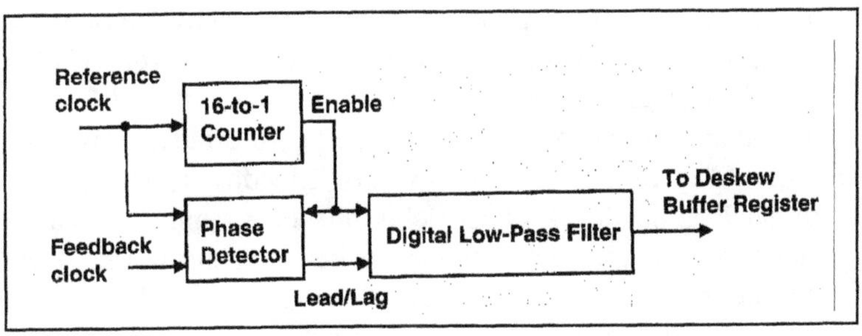

Figure 5-8. Control Logic for Delay Lines [83].

5.5 Clock Shrinking Technique for Silicon Debug

The post-silicon speed path debug has played an increasing role in the past few years. Without a good post-silicon debug methodology, it is difficult to increase the bin split through subsequent stepping beyond a few circuit modifications at a time. Techniques such as the PLL bypass mode have been used in the past. The PLL bypass mode intentionally bypasses the microprocessor's on-die PLL clock generator and relies on the external tester to supply the core clock. To apply this technique to a microprocessor that operates with frequencies exceeding 500MHz requires the automatic test equipment (ATE) tester to supply two out-of-phase clocks and an on-die clock mixer to generate the high frequency core clock [83]. Accurate core-clock edge manipulation cannot be achieved with this technique because of the limited edge placement accuracy (EPA) of the ATE clock signal. The EPA of a state-of-the-art ATE tester is approximately 100ps per clock channel [83]. When coupled to an on-die clock mixer and with two tester clocks, the core clock edge accuracy is 200ps. This accuracy is worthless when applied to the post-silicon timing debug in GHz microprocessors.

Another technique is to use the on-die clock shrink (ODCS) technique that can manipulate the clock edge at certain predetermined times. Figure 5-10 shows the circuits of the clock generation unit with the on-die clock shrink feature [83]. The on-die clock shrink block consists of the controller, an on-die clock shrink delay buffer, and the on-die clock shrink bus block generation circuitry. The controller contains three registers that will specify the rise and fall delays of the on-die clock shrink buffer in any three consecutive cycles. Since the on-die clock shrink mode manipulates the core clock edges and, in turn, affects the PLL feedback clock, there is the concern that the PLL may be affected. To ensure that the PLL clock generator is not affected by the on-die clock shrink operation, a dedicated PLL feedback path is utilized. The on-die clock shrink bus clock generation circuitry generates the PLL feedback clock that is unperturbed by the on-die clock shrink buffer. By inserting an additional delay to the on-die clock shrink feedback that corresponds to the normal PLL feedback clock delay, the FSB timing is unaffected. Any slight delay mismatch between the on-die clock shrink PLL feedback clock and the normal PLL feedback clock will not affect the PLL operation because the usage does not permit dynamic switching between these two modes.

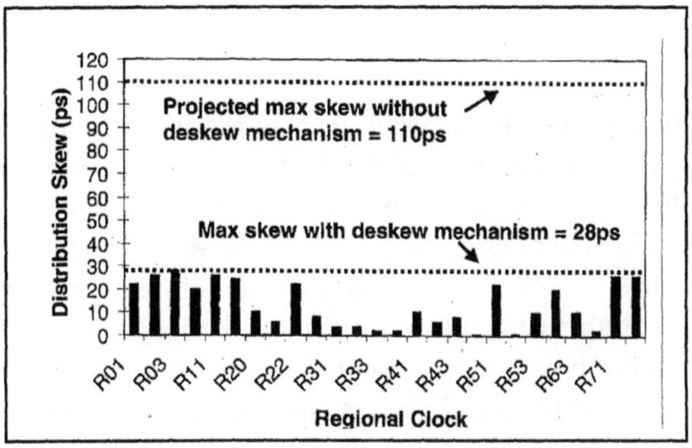

(a)

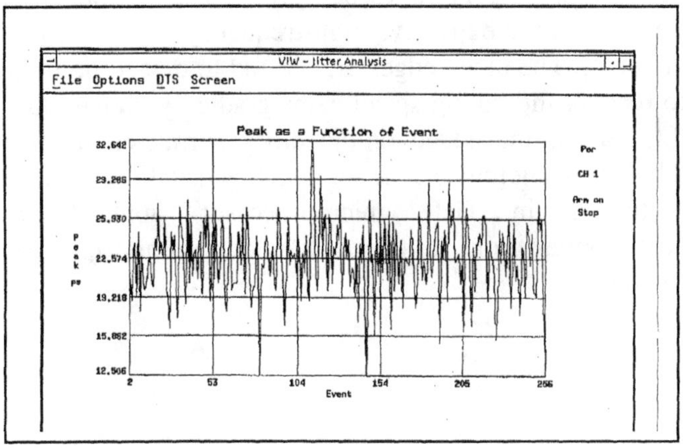

(b)

Figure 5-9. Skew and Jitter Measurement for Intel IA-64 De-skewing Design [83].

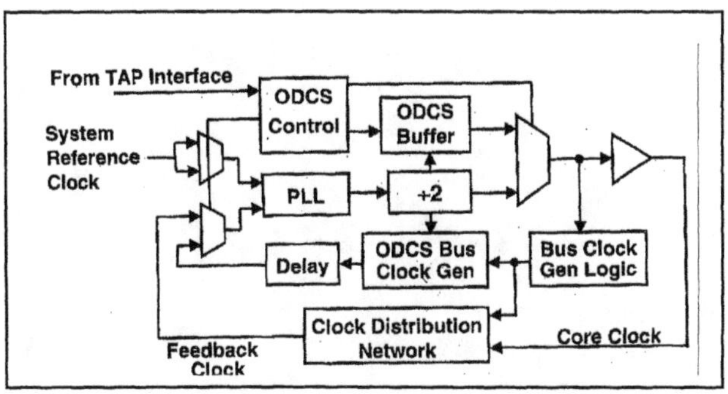

Figure 5-10: On-Die Clock Shrink [83].

5.6 Summary

The phase-locked loop circuit is critical for the generation of low-jitter clock signals for the chip with good phase alignment to the input clock. In addition, the frequency multiplications are usually provided by the PLL. The PLL block has to be placed in the entire chip with isolation to the digital cores. A set of PLL floorplan guidelines has been provided in this chapter. With faster clock frequency, larger die size, and multi-million transistors, clock skew control inside the chip is extremely difficult even with the balanced clock trees and balanced clock loads. The on-die PVT variations add to the difficulty in clock skew reduction. Therefore, the clock de-skewing technique (DLL) has been widely used in high-speed microprocessors in recent years. The on-die clock shrink technique can avoid the expensive high-frequency testers for post-silicon speed-path debug. The clock edges are shifted to detect the failing speed paths or to recover the failing speed paths gradually. Some jitter impacts on the PLL due to the added on-die clock shrink circuit need to be quantified before using this technique, which is usually a replacement for high-cost testers in GHz microprocessor applications. Other microprocessor applications for the on-die clock shrink techniques can be found in [64,86].

Chapter 6 Microprocessor Clock Distribution Examples

Aggressive clock skew control is always one of major efforts in the microprocessor design. Any clock skew reduction adds directly to the available cycle time. Many techniques have been explored in the clock design for high-performance microprocessors. Traditionally, the clock networks are routed in two hierarchies: the global clock network and the local clock networks. Balanced clock trees or clock grids are usually used to minimize the clock skew in the clock distribution. In recent years, the de-skewing buffers are added in the local clock network to further reduce the clock skew for the full-chip clock distribution. Section 6.1 shows the clock distribution scheme in an Intel IA-64 microprocessor design [83]. Section 6.2 presents the Intel Pentium IV clock distribution scheme [64]. Section 6.3 describes the Intel Pentium III clock distribution method [85,86]. Section 6.4 discusses the DEC Alpha chip clock distribution methodology [98]. Section 6.5 shows the IBM PowerPC clock distribution design considerations [101,102]. Section 6.6 contains a summary of this chapter.

6.1 Intel IA-64

The Intel IA-64 microprocessor clock distribution method and circuit techniques were reviewed in [83]. The overall clock distribution structure is in shown in Figure 5-4. It consists of the following three segments in the distribution path from the PLL to the clocked elements.

1. The global clock distribution includes the main clock tree and the reference clock tree from the PLL clock generator to the de-skewing buffers.

2. The regional clock distribution contains the clock grids that connect the clock from the de-skewing buffers to about 30 clock regions.

3. The local clock distribution contains the local clock buffers, which get the clock from the regional clock grid, and local interconnections to clocked elements.

The front-end global distribution consists of external clock input buffer, an on-die PLL, and the global clock distribution network. The microprocessor receives differential clocks running at the front-side-bus frequency. A high-performance PLL is used to generate a high-frequency clock running 2X the core clock frequency [96]. One input to the PLL is the system clock or the external clock. The other input to the PLL is the feedback clock, which is generated on-die and operates at the same frequency as the system clock and performs the front-side-bus clocking. Core frequency multiplication is accomplished with the on-die digital dividers acting on the high-frequency core clock. A clock input buffer converts the differential system clock into a single-ended clock with CMOS levels. An identical clock input buffer is inserted in the PLL feedback clock path to ensure that the FSB clock will be matched to the system clock for the bus transactions. Figure 6-1 shows the schematic of the differential input clock buffer, where clkn and clkp are two differential input clocks. This design shows a good common-mode range based on a fully complementary self-bias design [45].

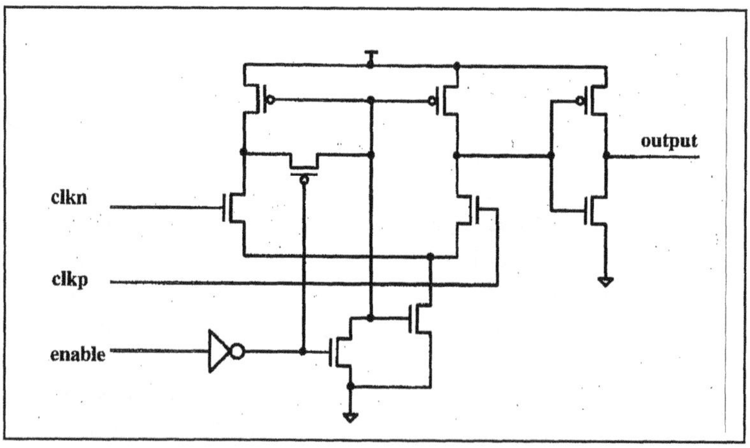

Figure 6-1: Differential Clock Input Buffer [83].

The core clock and reference clock are distributed from the PLL clock generator to de-skewing clusters through two identical and balanced H-trees. Figure 6-2 shows the global clock tree [83]. The tree is implemented exclusively in the top two metal layers. To reduce the capacitive noise coupling and to ensure good inductive return path, the global clock tree is completely shielded laterally with Vdd and Vss lines. In addition, the inductive reflections at the branch points are minimized by properly sizing the metal widths for impedance matching. A 2-D interconnect simulator computes the global clock tree inductance, capacitance, and the return path resistance for a distributed RLC optimization of the clock wire sizing. The reference clock tree has the same topology as the global clock tree. An intentional delay is added to the root of the reference clock. This additional delay is used to match the delays of the global clock path and reference clock path at the destinations of DSK buffers, as shown in Figure 6-2. The reduced loads in the reference clock tree allow the use of fewer or no intermediate buffers, as shown in the global clock tree, and thus reduce the clock skew due to PVT variations in clock buffers.

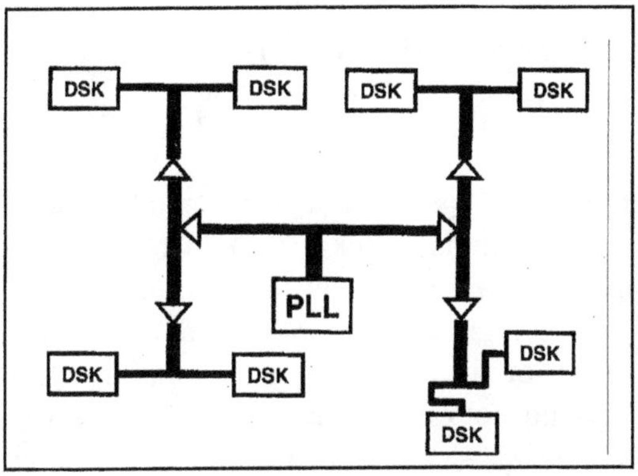

Figure 6-2: Global Clock Tree [83].

The regional clock distribution includes the de-skewing buffer, the regional clock driver, and the regional clock grid. The entire microprocessor chip consists of 30 separate clock regions, as shown in Figure 6-3. Every regional clock is mapped to a specific de-skewing buffer. The distributed de-skewing buffers are used to reduce the clock skew across the chip. For each clock region, a group of regional clock driver cells is used to drive the clock grid, which covers the clock region. The regional clock driver cell supports features like scan and clock gating.

A binary distribution network is used to connect de-skewing buffers to regional clock drivers. These binary routes are implemented on top metals with shielding of the Vdd and Vss lines. The regional clock drivers thus drive the clock grid for clock skew reduction across this clock region.

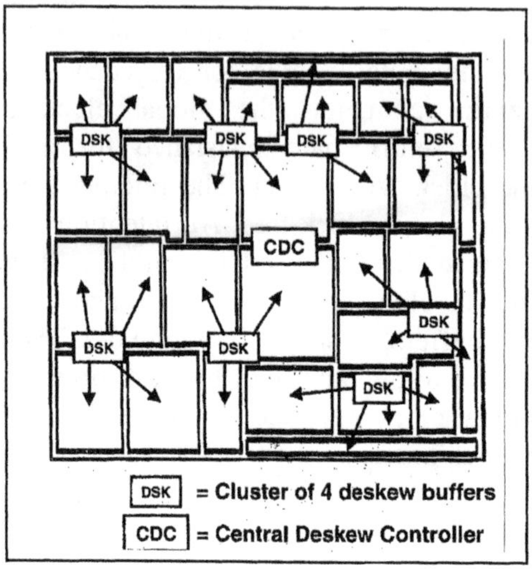

Figure 6-3: Regional Clock Distribution [83].

The regional clock grid is routed in metal 4 and metal 5. As the global clock tree, the regional clock grid is also laterally shielded using Vdd and Vss lines in order to reduce the coupling noise. The maximum dimension of a clock region is chosen based on the underlying clock loading and the span of the region. The regional clock driver cells are located on the top and bottom of the clock grid. The final stage of the clock network is the local clock distribution, which is the extended routing to clocked elements from the regional clock grid. It includes several stages of local clock buffers and local clock routings in a functional block. The global clock and regional clock distributions are the responsibility of the global clock owner, but the local clock is the responsibility of the functional block designers. It is a hierarchical design style for a microprocessor chip. The local clock design includes the insertion of the clock buffers and local clock routing. A family of local clock buffer cells is designed for a range of loads and for equal delays across all clock paths. For the time borrowing and domino circuits, additional local clock buffers may be inserted. Figure 6-4 shows the local clock distribution structure. The skew and clock jitter measurements for the clock distribution network in the IA-64 microprocessor can be found in Figure 5-9.

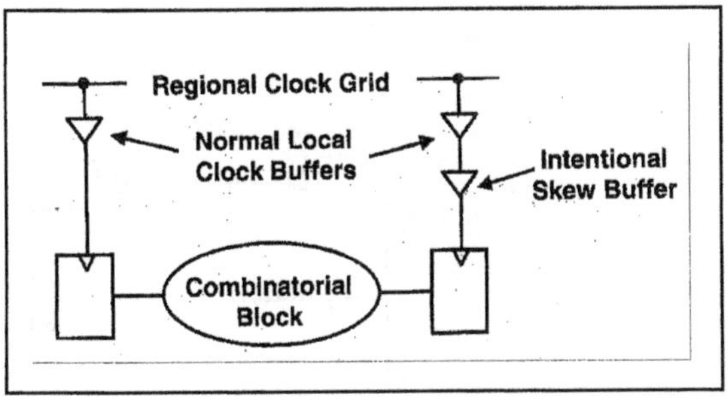

Figure 6-4: Local Clock Distribution [83].

6.2 Intel Pentium IV

A Pentium IV clock distribution technique operating at 2.0GHz has been reported in [64]. An external clock driver provides a differential low-swing system clock as the reference clock. Low-swing differential clocks are used to maximize the common-mode noise rejection. A two-stage double-differential receiver, shown in Figure 6-5, amplifies the low input swings to single-ended CMOS levels. The first stage is a low-gain, self-biased sense amplifier providing high tolerance to process, voltage, and temperature variations. The second stage is a high-gain stage used to drive clock out. This dual configuration provides a wide common-mode range and a high power supply rejection noise (PSRR), and enables small latency variations with changing operating and input conditions.

Two separate PLLs generate the core and I/O clocks in the Pentium-4 microprocessor [64]. A dedicated PLL for the I/O clock is used to de-couple I/O transactions rate from core frequencies while maintaining synchronous on- and off-chip operations. The core PLL feedback is generated using a programmable divide-by-N clock divider, located at the end of the clock network, thus enabling the core to operate at N times the system clock frequency. The I/O PLL feedback divides the I/O clock by a fixed factor of four, thus synthesizing an I/O clock at four times the system frequency for the data busses. The two PLLs provide a total of six different frequencies concurrently to different logic blocks in the die. Figure 6-6 shows different clock frequencies used in this microprocessor started from a 100MHz system clock. The core clocks range in 4GHz, 2GHz, 1GHz, 200MHz, and 100MHz. The I/O clock is in 400MHz.

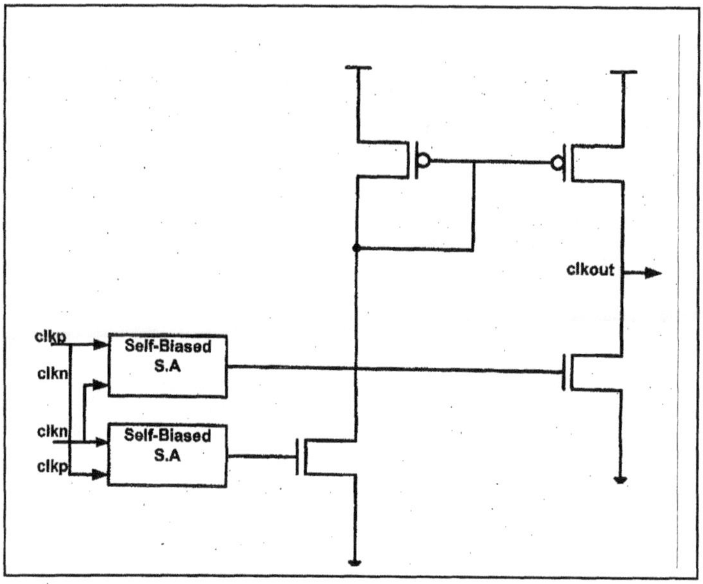

Figure 6-5: Differential Clock Input Buffer [64].

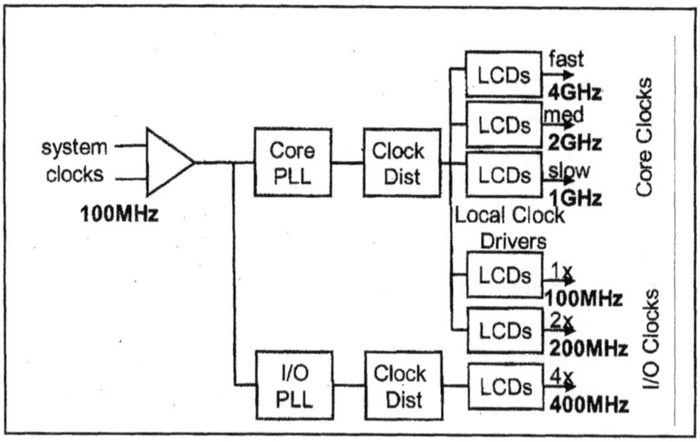

Figure 6-6: Clock Frequencies and Architecture [64].

The global clock tree includes three main clock spines, as shown in Figure 6-7(a). Each spine contains a binary distribution tree, with each leaf node providing an independent domain clock as shown in Figure 6-7(b). The wide sizing of the spine controls the clock jitter and clock skew to within 10% of the clock period. The power supply noise has a significant impact on the clock distribution jitters. A simple low-pass RC filter, as shown in Figure 6-8(a), is designed to reduce the core supply noise for clock buffers. The resistor R in Figure 6-8(a) is implemented in PMOS transistors. Design tradeoffs are made to accommodate the low IR drop with large RC time constant and at the same time minimize the layout area requirement for the capacitor. As shown in the inverter schematic in Figure 6-8(a), with 10% dip in the power supply line, the power supply for the clock buffer suffers about a 2% dip. The actual component values for the RC filter at different clock buffers are adjusted to maintain fixed IR drop and RC time constant. The simulation shows about 5X reduction in the noise amplitude on the filtered supply. Figure 6-8(b) shows the simulation results of cycle-by-cycle clock jitters through the clock buffer with and without the RC filter [64].

At the ends of the global clock distribution tree in Figure 6-7, there are 47 adjustable de-skewing buffers with phase detectors to minimize the clock skew. With the inclusion of the power supply filtering, the clock jitter due to the added latency of programmable de-skewing buffers is reduced. Figure 6-9 shows the skew reduction circuit using the de-skewing buffer (DB) before the clock is extended to the local clock distribution. The delay adjustment control for domain de-skewing buffers and the output of phase detectors are accessible from a test access port (TAP). With de-skewing buffers and adjustable delays, the clock skew is adjusted to within the accumulation error of about 8ps, limited mainly by the resolution of the de-skewing buffers [64].

Table 6-1 shows the clock jitter and clock skew results in the Pentium IV chip based on on-die measurements using the backside laser probing technique. The results are obtained based on all the de-skewing buffers and other balanced routing efforts in the clock distribution. The entire microprocessor measurements are made in a 0.18μm process in 4GHz pulsed clocks, and 2GHz and 1GHz pulsed and non-pulsed core clocks.

(a)

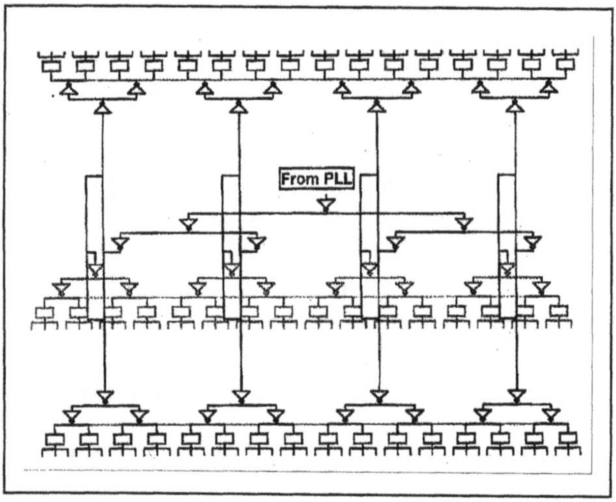

(b)

Figure 6-7: Global Clock Distribution [64].

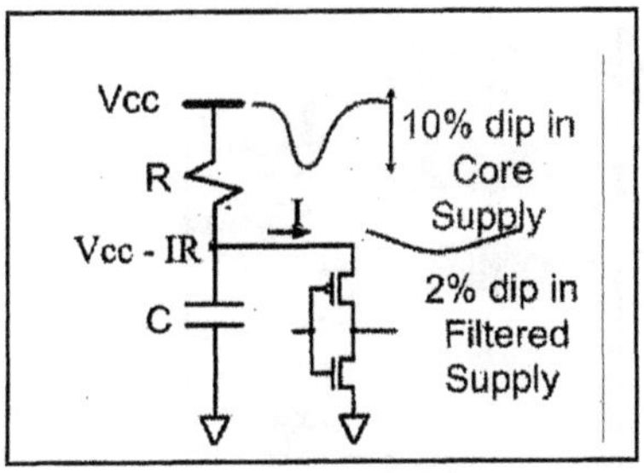

(a)

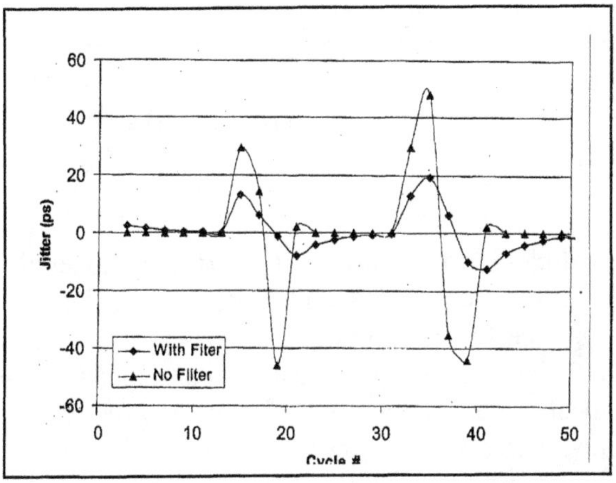

(b)

Figure 6-8: RC Filter for Power Noise Reduction [64].

Table 6-1: Skew and Jitter Measurements in Intel Pentium IV Chip [64].

Jitter (clock distribution)	Jitter (PLL)	Total jitter	Clock skew (worst)
30ps	5ps	35ps	16ps

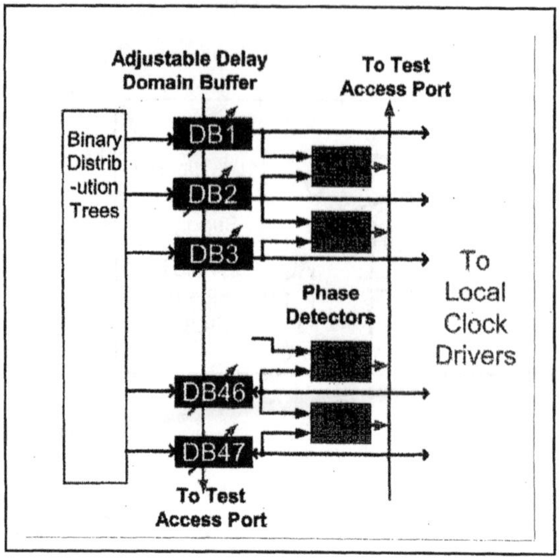

Figure 6-9: De-skewing Circuit [64].

The local clock drivers are used to drive the local clock load and to produce the clock frequencies and clock phases for each particular functional block. The local drivers are connected to appropriate domain clocks through delay-matched taps. The maximum RC delay from the output of the local drivers to the input of a latch is restricted to minimize local clock skew. The maximum local clock skew, including the RC line mismatches, is required to be less than 38ps in this microprocessor [64]. There are three types of local clock drivers, medium/fast/slow, and two types of clock signals, pulsed/non-pulsed regular clocks, in the design. The medium/fast/slow clock frequencies are 2GHz/4GHz/1GHz for different clock domains to feed different circuits in the core as shown in Figure 6-7. A non-pulsed regular clock driver only buffers the incoming global clock and is available at the slow and medium clock frequencies. Pulsed clock drivers are available at all clock frequencies. Pulses are generated from the rising, falling, or both edges of the global clock. The medium clock drivers that generate the medium clock frequency (2GHz) are used in most functional blocks of the core. The fast clock drivers that generate the fast clock frequency (4GHz) are used in the execution blocks, since execution blocks need the fastest performance. The slow clock drivers that generate the slow clock frequency (1GHz) are used in timing non-critical blocks.

6.3 Intel Pentium III

A two-spine global clock distribution structure is used in a 650MHz Pentium III chip, as shown in Figure 6-10 [85,86]. These two spines are properly shielded so that they will not be impacted with fringing fields from any interconnections in the core and I/O sides. The de-skewing circuit (DLL) reduces the clock skew between the spines and reduces clock jitter. The active de-skewing scheme helps to reduce the left-to-right clock spine skews from 100 to 15ps.

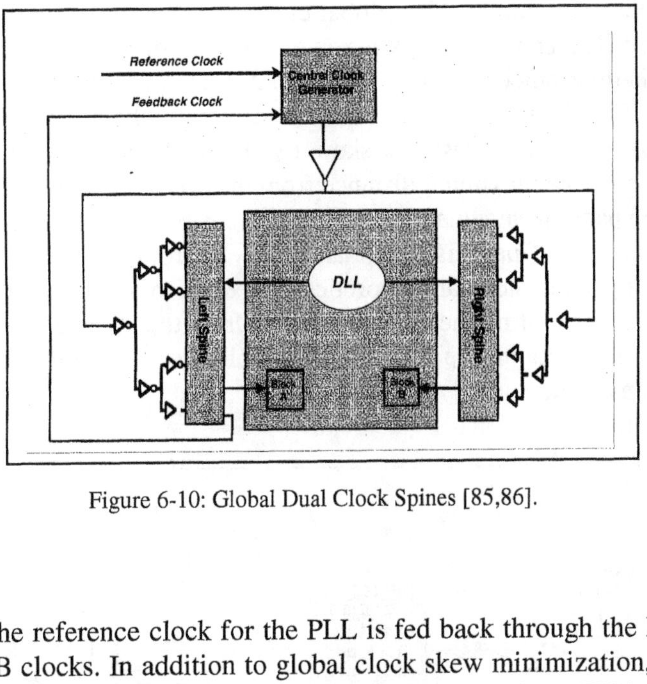

Figure 6-10: Global Dual Clock Spines [85,86].

The reference clock for the PLL is fed back through the left side of the FSB clocks. In addition to global clock skew minimization, careful driver and interconnect design matches the front-side bus (FSB) clock to the backside bus (BSB) clock. Both the backside bus and front-side bus clock spines are shielded to provide better isolation from cross talk. Bus clock skew minimization is essential for the operation of the FSB common clock up to 133MHz, and for full-frequency operation of the BSB using a source synchronous signalling scheme. In this processor design, several FSB-to-core clock ratios were implemented to support 66MHz, 100MHz, and 133MHz FSB operations while maintaining flexible processor core frequency choices. It allows the shut off of certain sections of the clock tree for sleep and deep-sleep modes for mobile applications.

6.4 DEC Alpha

The 600MHz Alpha chip's clock distribution network, shown in Figure 6-11, has been reported in [98]. A global clock grid (GCLK) is used in this design to reduce the clock skew for two-phase clocks. The clock generation is provided by an on-chip, low-jitter phase-locked loop (PLL) [90,97]. The PLL multiplies a low-frequency (80-200MHz) external clock and is used for I/O synchronization. It has a separate, regulated 3.3V power supply and is located in a corner of the chip to minimize the noise impact to the PLL. The clock distribution network up to and including GCLK is included in the feedback loop of the PLL to control phase alignment. The global clock distribution network is shown in Figure 6-12. The PLL clock is routed along a trunk to the center of the die and is distributed by the H trees and X trees to 16 distributed GCLK drivers. The arrangement of GCLK drivers, which resembles four "windowpanes," achieves low skew by dividing the chip into regions, thus reducing the maximum distance from the drivers to the farthest loads. A windowpane arrangement also reduces sensitivity to process variation because each grid pane is redundantly driven from four sides, although only two opposite sides are theoretically needed to attain the same skew. In general, distributing the drivers widely across the chip also has the dual benefits of reducing power supply collapse and improving heat dissipation efficiency.

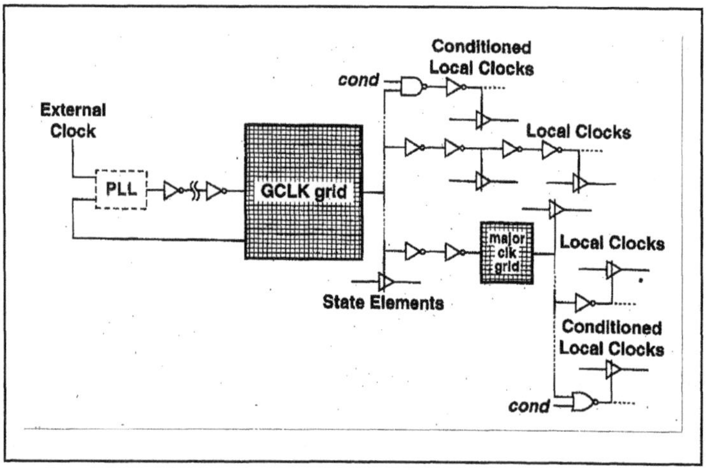

Figure 6-11: Clock Distribution Hierarchy in DEC Alpha Chip [98].

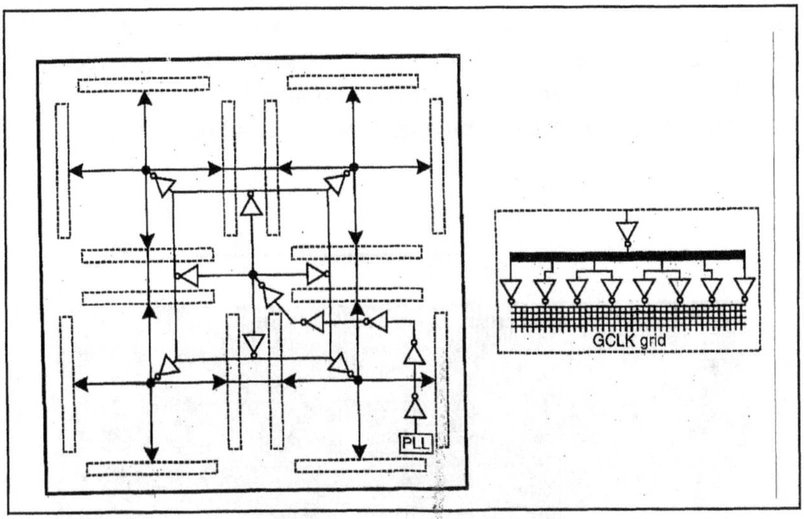

Figure 6-12: Global Clock Network [98].

The final two stages of the GCLK distribution network, as shown in the dashed boxes in Figure 6-12, use a balanced binary tree. The binary tree shown on the left side equalizes the delays from the central pre-driver to all the individual GCLK drivers. Since the GCLK drivers are large, the Elmore delay model is dominated by the interconnect resistance times the capacitance of the receiving gate. For the clock distribution balancing purpose, the interconnect resistance of an RC tree is less than half the resistance of a tapered H tree for the same number of metal tracks. Consequently, in this case, the pre-driver for the tapered H tree would have to be 50% larger than that for the RC tree in order to achieve the same rise and fall times. The smaller size of the pre-driver for the RC tree therefore allows preceding driver sizes to be also reduced, thus compounding the power and area saving.

The primary disadvantage of a girded clock is the extra capacitance of the grid. At 600MHz and 2.2V, the typical power usage for the processor is 72W, the global clock (GCLK) uses 10.2W, and the complete distribution network that eventually drives GCLK uses 5.8W. With a gridded clock, the trade off between power and performance is essentially determined by the choice of the skew target , which determines the needed grid density and therefore the clock driver sizes [98]. Figure 6-13 shows the layout of the GCLK grid in the Alpha microprocessor [98]. It traverses the entire die and uses 3% of the upper level low-impedance interconnect, i.e., Metal 3 and Metal 4. All the clock interconnects are laterally shielded with either Vdd and Vss interconnects. The microprocessor also has a Vss reference plane sandwiched between Metal 2 and Metal 3 and a Vdd reference plane above Metal 4. All the clock wires and all lateral shields are manually designed in this chip [98].

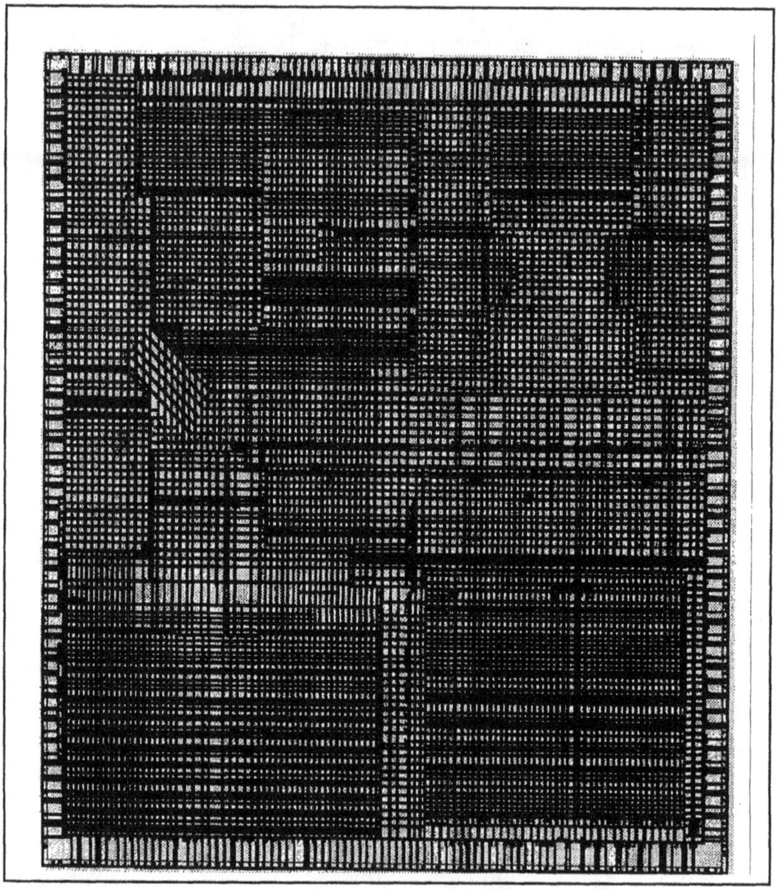

Figure 6-13: Global Clock Grid [98].

After the global clock grid (GCLK) with two more global clock drivers, there are several regional clock grids over their respective execution units, as shown in Figure 6-14. There are in total six regional clock grids (CCLK, ECLK, FCLK, JCLK, MCLK, and PCLK) in this chip. The empty areas in Figure 6-14 are exclusively used for local clocks, although the local clocks can also be found in the regional clock grid areas. The grid density varies widely among regional clocks and even within a single regional clock. The variation in regional clock grid density reflects clock loads. The regional clock grids are sized to meet the clock skew targets. At 600MHz and 2.2V supply, the regional clocks use about 14.0W.

The design of local clock distribution in the Alpha microprocessors is not gridded and not shielded. The design guidelines also show no strict limits on the number, size, or logic function of local clock buffers, although timing path constraints must always be satisfied. There are requirements for the rise and fall times for the local clocks on the clocked elements. Gated clocks are used to reduce the power. At 600MHz and 2.2V, local unconditional clocks use 7.6W and local conditional clocks use a maximum of 15.6W, assuming they switch every cycle [98].

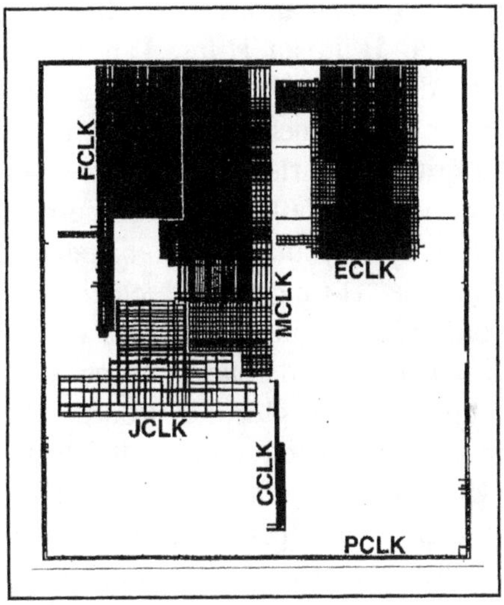

Figure 6-14: Regional Clock Grids [98].

6.5 IBM PowerPC

A high performance microprocessor PowerPC 603 clock distribution structure and PLL circuit have been reported in [89]. An H-tree style, balanced clock tree has been used in a series of PowerPC microprocessors [89,101]. Given a set of metal lines of width W and length L, the delay of a signal starting at the middle of a clock distribution wire in the shape of an H to the top and bottom endpoints of the H is electrically the same. Figure 6-15 shows the clock distribution in PowerPC microprocessor based on an H-tree, which replicates as many H's as are necessary to cover the entire clock distribution area. This is done in a uniform fashion extending from the endpoints of each preceding H. The design priorities for the clock distribution are minimum clock skew, sharp rise and fall times, 50% duty-cycle, and low power consumption. Rise and fall times below 100ps are typically the design target for a 1ns CPU clock period.

The H-tree with the combination of the clock grid structure has been extensively discussed in [102]. The methodology is as follows. The initial stage of designing the global clock distribution and the clock driver requires an estimate of the total number of loads on the H-tree, their input capacitance, and an approximate total area of coverage. This allows the designer to perform SPICE simulations with RC network representations for each branch of each H in an H-tree. These RC networks or pi-networks have process dependent R and C with embedded variations for width, thickness, and metal-to-metal spacing. The clock driver size is chosen based on the criteria for rise and fall times and minimum skew. The driver delay and rise and fall times are characterized with proper parasitic extracted from the layout. An automated routing tool connects the last branches of the H-tree to the clock inputs of the clock regenerators. A layout-based timing analysis is performed to verify the skew control targets. The measured delay is 230ps. A total of 135ps of the clock skew at the inputs of the global H-tree is measured. The clock regenerators are the de-skewing circuits for the local clock buffers, as shown in Figure 6-15. The clock distribution for 266 clock regenerators with 80fF of input capacitance per regenerator dissipates 120mW at 80MHz including the PLL and clock driver circuits.

With the goal of developing a single global clock distribution method to be used in a variety of high-performance server microprocessors including IBM PowerPC, a strategy consisting of many trees driving a single clock grid has been developed [101]. This method is applied to several chips. The global clock network consists of a series of buffered tree-like networks (H-trees) driving a final set of 16-64 sector buffers. While the number and placement of these sector buffers vary greatly among chips, in all cases each buffer drives a final tuneable tree network. All of these tuneable trees drive the single clock grid covering the die area. Figure 6-16(a) shows the global clock distribution structure in a test chip [101]. This distribution network topology consists of a number of tuneable H-trees driving a common clock grid. The clock grid reduces the local skew by connecting the nearby points directly. The tree wires are tuned to minimize the clock skew over long distances.

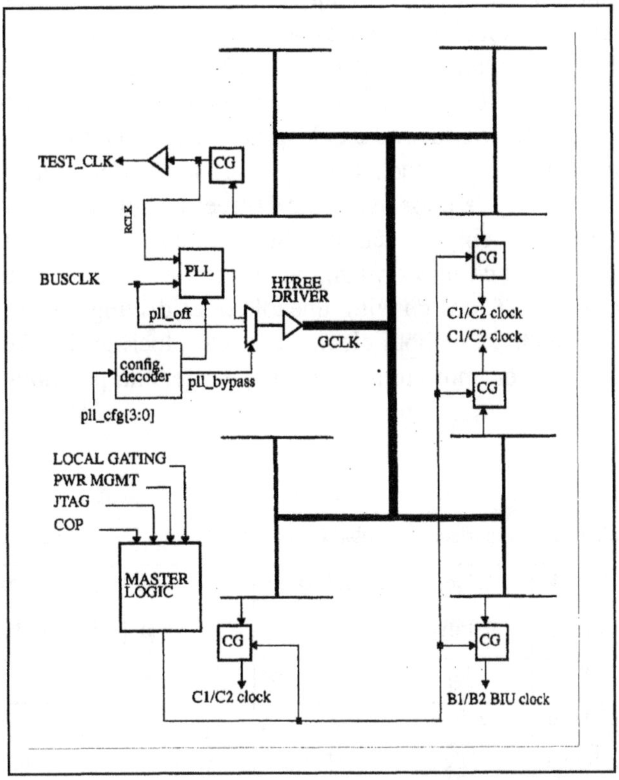

Figure 6-15: Global Clock Distribution in H-Tree Style in IBM PowerPC [98].

6.6 Summary

Three basic routing structures are used for the global clock network in the leading microprocessors. The H-tree clock tree method provides the lowest clock skew and the smallest routing capacitance. It has the benefit of reducing power consumption. The clock grid can tie the elements together and, to some extent, reduce clock skew. But the clock grid increases the routing capacitance for the clock distribution network, and has worse power consumption. The Alpha microprocessor chip design style uses the global clock grid and regional clock grids. The clock spine has very small RC delay because of the wide spine width. Usually one center clock spine or several spines are used across the chip. But the clock spine method has to balance the delays to the locations in the chip and usually that still causes significant skew. The routing capacitance is lower than that of the clock grid but usually higher than that of the H-tree structure. Table 6-2 summarizes the trade off between clock skew and power consumption for three common global clock distribution topologies: H-tree, clock grid, and clock spine. Obviously, the H-tree style is a winner according to the clock skew, power, and layout area, although the floorplan flexibility is poor using the H-tree design. The H-tree goes everywhere across the chip to achieve the recursive balanced structure. For changes in the full chip floorplan, the H-tree has to be laid out again. The method in PowerPC design with tuneable trees driving the clock grids combines the advantages of both clock trees and clock grids. Trees have low latency, low power, minimum wire track usage, and potential for low skew.

Table 6-2: Comparison of Global Clock Distribution Topologies.

Clock Structure	**Clock Skew**	**Capacitance**	**Layout Area**	**Floorplan Flexibility**
H-Tree	Low	Low	Low	Low
Clock Grid	Medium	High	High	High
Clock Spine	High	Medium	Medium	Medium

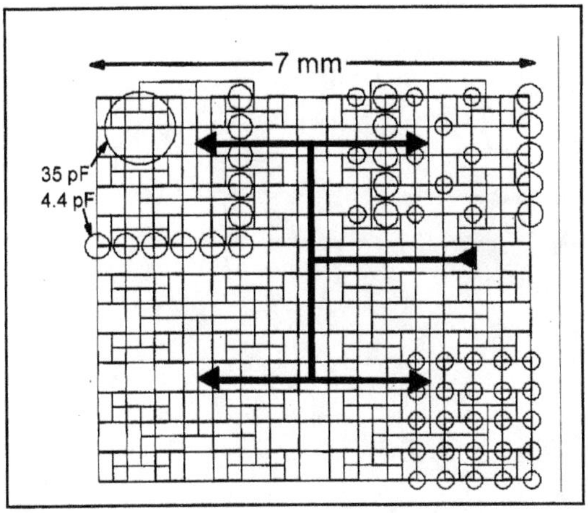

(a)

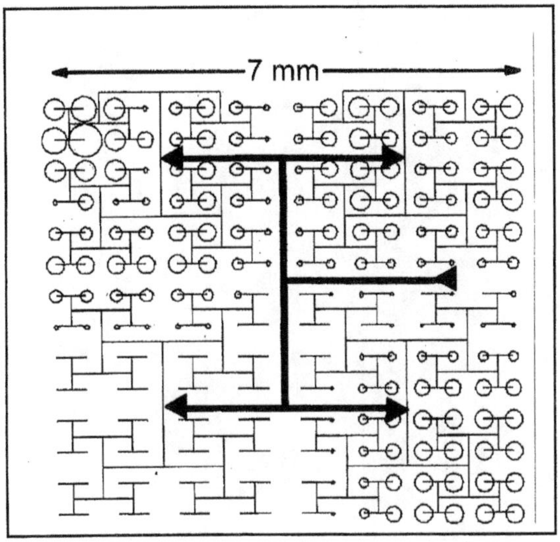

(b)

Figure 6-16: H-trees Driving Embedded Grid [101].

Chapter 7 Clock Network Simulation Methods

Because of the importance of the clock circuit, designers have been verifying in detail the clock networks for microprocessor chips and ASIC chips [99,107]. The manual approach of simulating each stage of the clock buffer tree has been widely used in the design of most chips, although the automation CAD tools for the entire clock network analysis are available nowadays [107,112]. The manual approach usually consists of the following procedures for the clock tree analysis: (a) identify the entire clock tree; (b) separate the clock network into the stages; (c) extract RC data of the clock lines at each stage; (d) simulate each stage of the clock tree including the RC line models and buffer models; and (e) piece the simulation results together to trace the clock delays from the root to the end points. The delay for each stage is measured at the inputs of the buffers to the inputs of the next stage. The above approach may not be feasible for the clock mesh structure such as in the Alpha microprocessor chip [107]. Section 7.1 introduces the RC extraction flow for clock network. Section 7.2 demonstrates the clock tree tracing and RC stitching capability by a CAD tool [112]. Section 7.3 shows the simulation methods and report files for clock networks. Section 7.4 discusses the effects of the IR drop in the power supply network on the clock skew analysis. Section 7.5 gives the summary to this chapter.

7.1 RC Extraction

All the clock skew analysis methodologies require the complete RC data extracted from the layout for the modelling of full-chip clock networks. Figure 7-1 illustrates the generic RC parasitic concepts in the circuit model. The distributed RC parasitics will include the metal resistance in the model; while the lumped capacitive parasitics only include the capacitance extracted from the layout interconnects. Figure 7-2 illustrates the distributed RC elements between clock buffers (G1, G2, etc.) in the full-chip clock tree [107]. These extracted clock buffer stages, together with RC elements and transistor models, are used in the clock tree simulation.

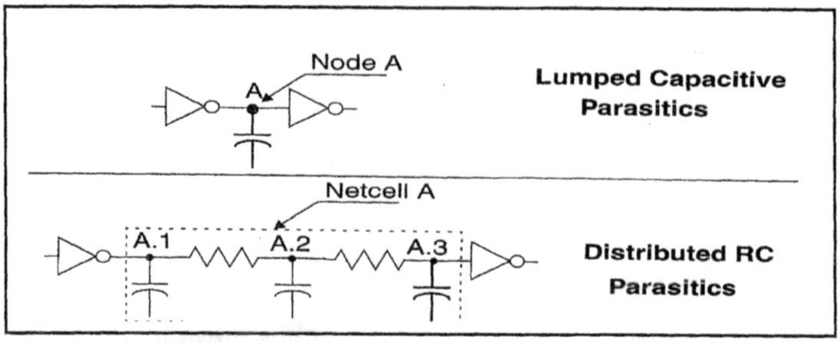

Figure 7-1. RC Models.

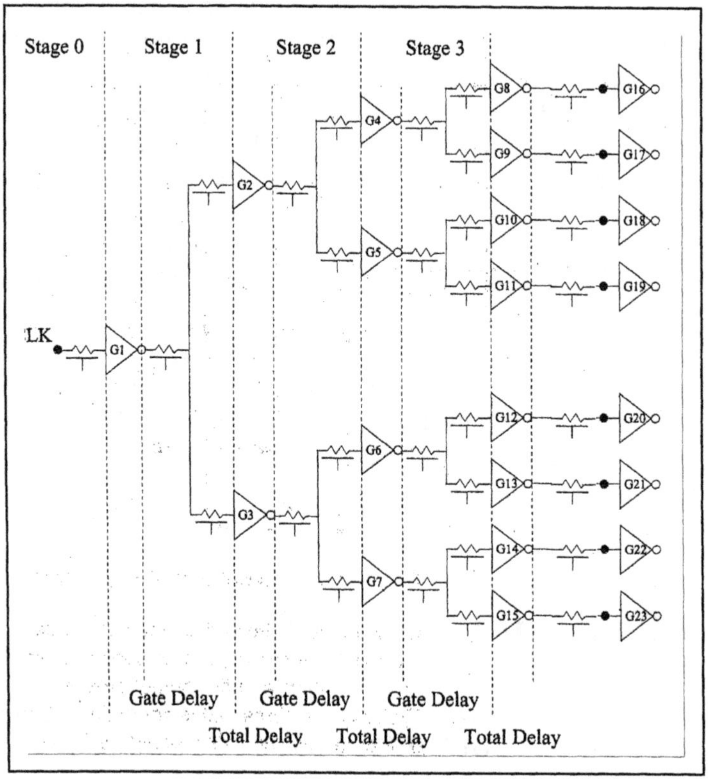

Figure 7-2. RC Modelling of Clock Distribution Tree [107].

The RC extraction works on a physical layout database to extract resistance and capacitance models. The modelling procedure from the layout is illustrated in Figure 7-3. The resistance and capacitance models are stitched back to the specific circuit simulation netlist in the pre-layout design stage. The resistance model is calculated by the following well-known equation:

$$R = \rho l / w \ (ohm). \tag{7-1}$$

Where ρ is the sheet resistance in ohm/square unit, l is the length of the line in m unit, and w is the width of the line in m unit. Table 7-1 shows the approximate sheet resistance in a 0.18μm technology. The capacitance model is much more complicated, since the capacitances from a metal line to adjacent lines are composed of several components, as illustrated in Figure 7-4. The capacitance models are usually built up from the equations for the variables of the geometrical distances from one metal line to neighbouring lines.

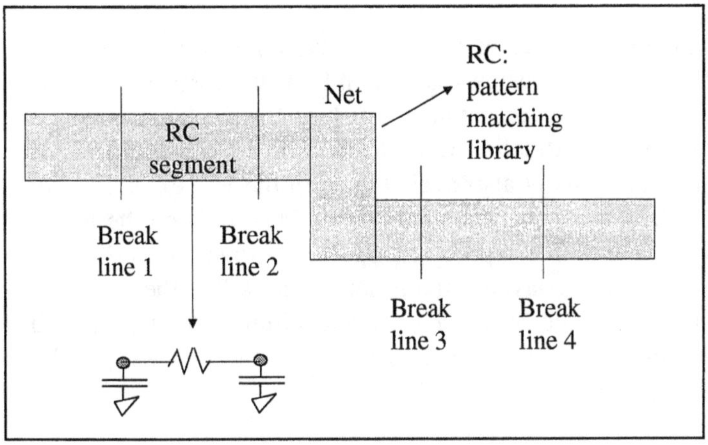

Figure 7-3. RC Modelling of Metal Segment.

Table 7-1. Sheet Resistances in a 0.18μm Technology.

Layer	P diffusion	N diffusion	Polysilicon	M1	M2	M3	M4	M5
ρ **(Ω/square)**	5.5	5.5	5.5	0.1	0.05	0.05	0.01	0.01

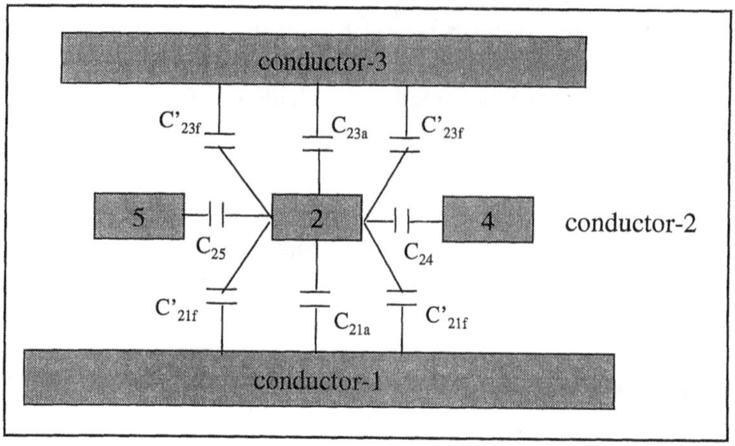

Figure 7-4 Interconnect Capacitance Components.

The detailed RC extraction results are saved in a standard parasitic format (SPF) file. Figure 7-5 shows an example of the SPF file, which includes the resistance R and capacitance C values for the layout metal segments based on the RC extraction CAD tools. The coupling capacitance report between metal lines is also available from the RC extraction tool [112]. Table 7-2 lists the most commonly used CAD tools on the RC extraction and most of them will support the SPF file format. This SPF file will be read back to the pre-layout circuit netlist and stitch the RC data with the device models to construct a complete circuit simulation netlist. This circuit simulation netlist is used to analyse the circuit with the RC effects to model the layout.

```
*|NET MECOPIPE0_ANI_FILTER_pix1_anichan0R_0 0.01518pf
*|I (MECOPIPE0_ANI_FILTER_U894:D2 MECOPIPE0_ANI_FILTER_U894 D2 I 0.00490pf
0.0 0.0 )
*|I (MECOPIPE0_ANI_FILTER_pix1_anichan0rR_reg_0:D
MECOPIPE0_ANI_FILTER_pix1_anichan0rR_reg_0 D I 0.00270pf  0.0 0.0 )
*|I (MECOPIPE0_ANI_FILTER_pix1_anichan0R_i_reg_0:O
MECOPIPE0_ANI_FILTER_pix1_anichan0R_i_reg_0 O O 0.00340pf  0.0 0.0 )
*|I (MECOPIPE0_ANI_FILTER_U994:A MECOPIPE0_ANI_FILTER_U994 A I 0.00470pf  0.0
0.0 )
*|S ( MECOPIPE0_ANI_FILTER_pix1_anichan0R_0:24 799.680 558.360 )
*|S ( MECOPIPE0_ANI_FILTER_pix1_anichan0R_0:21 799.840 558.360 )
*|S ( MECOPIPE0_ANI_FILTER_pix1_anichan0R_0:20 799.280 558.360 )
*|S ( MECOPIPE0_ANI_FILTER_pix1_anichan0R_0:29 822.720 577.080 )
R1 MECOPIPE0_ANI_FILTER_pix1_anichan0R_i_reg_0:O
MECOPIPE0_ANI_FILTER_U994:A 1.975917E+01
```

```
R2 MECOPIPE0_ANI_FILTER_pix1_anichan0R_0:24
MECOPIPE0_ANI_FILTER_pix1_anichan0R_0:21 3.495291E-02
R3 MECOPIPE0_ANI_FILTER_pix1_anichan0R_0:24
MECOPIPE0_ANI_FILTER_pix1_anichan0R_0:20 3.495291E-02
R4 MECOPIPE0_ANI_FILTER_U894:D2 MECOPIPE0_ANI_FILTER_pix1_anichan0R_0:20
9.646808E+00
C1 MECOPIPE0_ANI_FILTER_U894:D2 VSS 0.00177pf
C2 MECOPIPE0_ANI_FILTER_pix1_anichan0rR_reg_0:D VSS 0.00015pf
C3 MECOPIPE0_ANI_FILTER_pix1_anichan0R_i_reg_0:O VSS 0.00034pf
C4 MECOPIPE0_ANI_FILTER_pix1_anichan0R_i_reg_0:O VSS 0.00213pf
C5 MECOPIPE0_ANI_FILTER_U994:A VSS 0.00203pf
C6 MECOPIPE0_ANI_FILTER_pix1_anichan0R_0:20 VSS 0.00163pf
C7 MECOPIPE0_ANI_FILTER_pix1_anichan0R_0:21 VSS 0.00322pf
C8 MECOPIPE0_ANI_FILTER_pix1_anichan0R_0:24 VSS 0.00005pf
C9 MECOPIPE0_ANI_FILTER_pix1_anichan0R_0:29 VSS 0.00386pf
*
*|NET MECOPIPE1_ANI_FILTER_pix1_anichan0R_0 0.03250pf
*|I (MECOPIPE1_ANI_FILTER_U841:D2 MECOPIPE1_ANI_FILTER_U841 D2 I 0.00490pf
0.0 0.0 )
*|I (MECOPIPE1_ANI_FILTER_pix1_anichan0rR_reg_0:D
MECOPIPE1_ANI_FILTER_pix1_anichan0rR_reg_0 D I 0.00270pf  0.0 0.0 )
*|S ( MECOPIPE1_ANI_FILTER_pix1_anichan0R_0:28 1208.640 1045.080 )
*|S ( MECOPIPE1_ANI_FILTER_pix1_anichan0R_0:29 1210.080 1048.680 )
R1 MECOPIPE1_ANI_FILTER_pix1_anichan0R_0:28 MECOPIPE1_ANI_FILTER_U995:A
7.406403E+00
R2 MECOPIPE1_ANI_FILTER_pix1_anichan0rR_reg_0:D
MECOPIPE1_ANI_FILTER_U841:D2 1.553266E+01
R3 MECOPIPE1_ANI_FILTER_pix1_anichan0R_0:29 MECOPIPE1_ANI_FILTER_U995:A
5.850000E+00
R4 MECOPIPE1_ANI_FILTER_pix1_anichan0R_0:29
MECOPIPE1_ANI_FILTER_pix1_anichan0R_i_reg_0:O 1.815539E+01
R5 MECOPIPE1_ANI_FILTER_pix1_anichan0R_0:28
MECOPIPE1_ANI_FILTER_pix1_anichan0rR_reg_0:D 4.481287E+01
C1 MECOPIPE1_ANI_FILTER_U841:D2 VSS 0.00162pf
C2 MECOPIPE1_ANI_FILTER_pix1_anichan0rR_reg_0:D VSS 0.01238pf
C3 MECOPIPE1_ANI_FILTER_pix1_anichan0rR_reg_0:D VSS 0.00160pf
C4 MECOPIPE1_ANI_FILTER_U995:A VSS 0.00031pf
C5 MECOPIPE1_ANI_FILTER_U995:A VSS 0.00013pf
C6 MECOPIPE1_ANI_FILTER_pix1_anichan0R_i_reg_0:O VSS 0.00162pf
C7 MECOPIPE1_ANI_FILTER_pix1_anichan0R_0:28 VSS 0.01289pf
C8 MECOPIPE1_ANI_FILTER_pix1_anichan0R_0:29 VSS 0.00193pf
```

Figure 7-5. Standard Parasitic Format (SPF) File.

Table 7-2. RC Extraction CAD Tools.

Tool	**Company**
Fire-ice	Simplex Solution
Star-RC	Advanti!
Xcalibre	Mentor Graphics
HyperExtract	Cadence
Arcadia	Synopsys
Columbus	Frequency Technology
Nautilus	Ultima Interconnect
QuickCap	Random Logic

7.2 Full-Chip Clock Tree Tracing

The clock tree is traced from the phase-locked loop (PLL) to the clocked elements by the automation tools [107,112]. Figure 7-6 shows a clock tree traced by the StarRC-XT tool [112]. The clock network is traced forward from the start net named in the CLOCK_INPUT_SOURCE command, and ends at the input pins of any cell on the CLOCK_STOP_CELLS list. If the start net is a top-level port, the traced network begins at, and includes, the port. Alternatively, if the start net is an intermediate net, the traced network begins at the nearest instance outputs and includes the star net.

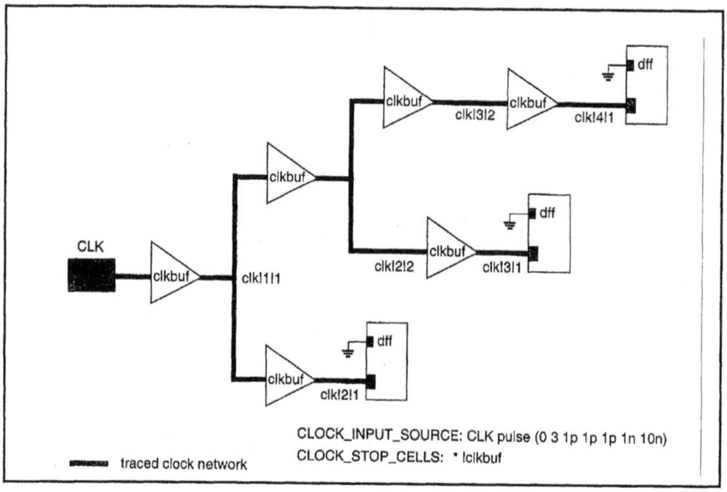

Figure 7-6. Clock Tree Tracing [112].

7.3 Clock Tree Simulation and Report Files

After obtaining the RC data and stitching the RC data back to the entire clock network by the clock tree tracing procedure, the next task will simulate the entire clock network and reports the clock delays and skew. The Star-RCXT tool pre-processes the simulation inputs to maximize the efficiency and measure the arrival times and edge rates of the clock tree [112]. Star-RCXT feeds only those spice sub-circuit models that participate in the clock network to the specified clock simulator, so there is no need to filter the cell library manually. Since Star-RCXT controls the simulation, the parasitic netlist is ensured to be compatible with the SUBCKT models. Each branch of the clock tree is simulated independently, reducing the capacity required to complete the simulation. The clock input source is applied to the network differently depending on how the network is traced. If the root net is a top-level port, then the voltage source is applied directly to that port. If the root net is an intermediate net, the voltage source is applied directly at the input of the previous stage.

Star-RCXT writes out the parasitic netlist internally. The special clock network format uses an integer-based naming scheme; this eliminates the need for extensive and messy name alterations to achieve SPICE compatibility. Each signal in the netlist is mapped to its original schematic name for all output and analysis functions within Star-RCXT. This clock netlist includes *|I and *|NET statements with schematic names for simple cross-referencing. The spice instrumentation file written by Star-RCXT is also annotated with schematic names for easy reference. In addition, Star-RCXT Clock Analysis provides four ways for the timing information: (1) clock tree browser; (2) SDF netlist; (3) ASCII and skew reports; and (4) simulation output. The clock tree browser is a tool with an expandable tree browser annotated with the simulation results and features a push-button layout that allows cross probing of every signal in the clock network. An SDF file records the RC parasitic models for the clock network by tracing the clock tree from the PLL. In addition, the Star-RCXT clock analysis generates two simple text files that detail the extract simulation output for arrival times and edge rates for each instance pin. The ASCII output report is organized by stages, sorted by the rising delay and contains the result of every measurement taken during the simulation. The SKEW report is similar to the ASCII report but only prints the timing information at each of the registers. The SKEW report also contains a short summary that identifies the minimum and maximum paths.

Table 7-3 shows the clock setup command options. In the CLOCK_INPUT_SOURCE command, the first argument is the root net, and must be a valid schematic name. The second argument must be a valid SPICE waveform. The clock paths are traced down from the specified root net, can begin from any level of the hierarchy, and continue on the gate level through the timing arcs from the library cell file. This command may be specified multiple times to extract and simulate more than one clock tree in a single job. Each clock network will be simulated and analyzed independently. The clock network is traced down from the

CLOCK_INPUT_SOURCE. When a cell matching the pattern in CLOCK_STOP_CELLS is encountered during tracing, every instance of the cell becomes a stopping point, which will be evaluated as a register. Using the CLOCK_STOP_CELLS, it is possible to identify all logic, register, and memory cells as the end points of the clock tree. Star-RCXT uses the loading capacitances for the CLOCK_STOP_CELLS and CLOCK_STOP_INSTANCES, extracted from the input layout database or the SYNOPSYS timing library, to create a simplified .SUBCKT model for the cell. The simple .SUBCKT model contains only capacitors to approximate the loading effects of the cell. The CLOCK_STOPCELLS_SUBCKT: YES option may be used to override Star-RCXT's model substitution. This will simulate the full clock network with the .SUBCKT models for the stop cells that were provided with the SPICE_SUBCKT_FILE.

Table 7-3. Clock Setup Commands [112].

Command	**Syntax**	**Description**	**Default Value**
CLOCK_INPUT_ SOURCE	CLOCK_INPUT_ SOURCE: <root net> <SPICE waveform>	The root net and input driving voltage for the clock network.	None
CLOCK_STOP_ CELLS	CLOCK_STOP_ CELLS: <cell>	A list of cell names for clock network tracking from the root to be stopped.	None
CLOCK_STOP_ INSTANCE	CLOCK_STOP_ INSTANCE: <instance>	A list of instance names for clock network tracing to be stopped.	None
CLOCK_ STOPCELLS_ SUBCKT	CLOCK_ STOPCELLS_ SUBCKT: <YES | NO>	YES will use the SPICE_SUBCKT_FILE S for the STOP_CELLS in clock analysis.	No
SPICE_CONTROL_ FILE	SPICE_CONTROL_ FILE: <file>	Simulation control files provided from user, which include the device models and simulator options.	None
SPICE_SUBCKT_ FILE	SPICE_SUBCKT_FI LE: <file>	.subckt definitions for all SKIP_CELLS.	None
SYNOPSYS_LIB_ FILE	SYNOPSYS_LIB_ FILE: <file>	The path to Synopsys .lib timing database.	None

Table 7-4 shows the clock simulation command options. The NET_VOLTAGE option defines a voltage to be used during the clock tree simulation, which can be used to sensitize any net in the clock network. The net name is any valid schematic name. The <voltage> can be any valid SPICE syntax. This command should be used for all power nets and any inputs to the clock tree such as the gated signal to the gated clocks. A warning is generated for any nets that do not have defined voltages and they are grounded prior to the simulation. This command can be specified many times, but there is only one entry per line. The PORT_VOLTAGE is applied to every instance of a cell and is useful for setting the data pins and the enable pins for latches.

Table 7-4. Clock Simulation Commands [112].

Command	Syntax	Description	Default Value
CLOCK_CROSS	CLOCK_CROSS: <Integer>	Clock period that the delays and edge rates are measured.	1 (first period)
CLOCK_DELAY_ THRESHOLD	CLOCK_DELAY_ THRESHOLD: < Float between 0 and 1>	Percentage of (V_{dd}-V_{ss}) for the trigger point in delay measurement.	0.5
CLOCK_HIGH_ THRESHOLD	CLOCK_HIGH_ THRESHOLD: < Float between 0 and 1>	Percentage of (V_{dd}-V_{ss}) for the end point in rising time measurement.	0.9
CLOCK_LOW_ THRESHOLD	CLOCK_LOW_ THRESHOLD: < Float between 0 and 1>	Percentage of (V_{dd}-V_{ss}) for the starting point in rising time.	0.1
CLOCK_ SIMULATION_ STEP	CLOCK_ SIMULATION_ STEP: < Float >	Time step used in clock simulator transient analysis for accuracy control.	Simulation time divided by 1000
CLOCK_ SIMULATION_ TIME	CLOCK_ SIMULATION_ TIME: < time>	Length of the clock simulation.	None
CLOCK_ SIMULATOR	CLOCK_SIMULATOR: < Star_simxt | Hspice >	Simulator to be used.	Star-Simxt
CLOCK_ WAVEFORM	CLOCK_WAVEFORM: < YES | NO >	Whether or not the clock waveform files to be saved.	No
NET_VOLTAGE	NET_VOLTAGE: < net > < voltage >	Voltage source at power nets and clock inputs.	None
PORT_VOLTAGE	PORT_VOLTAGE: < cell > <port> <volt>	Voltages for ports of a cell.	None

Table 7-5 shows the clock output command options. The CLOCK_SDF_OUTPUT specifies that the tool will generate the SDF file for the clock network, which provides, in standard timing format, the detailed timing data back-annotated to the logic synthesis tool for the cell sizing and optimization in the logic synthesis stage. Figure 7-7 shows the SDF file from the Clock Analysis [112]. It includes the delays of the interconnect portions and the cell portions, which have three components of the delay in the fast, typical, and slow simulation corners. The CLOCK_SKEW_REPORT_FILE will generate a summary file that lists the following information. Figure 7-8 shows the skew summary file from the Star-RCXT on the clock tree, which includes the following information.

- Arrival times and edge rates at each CLOCK_STOP_CELL instance by stages.
- The maximum clock skew in the clock distribution.
- The longest delay in the clock distribution or the latest arrival time to the CLOCK_STOP_CELL.

One example of the microprocessor clock network analysis can be found [98]. To simulate the global clock (GCLK) grid, an RC equivalent model is extracted from the layout of the GCLK grid and gate loads on GCLK [98]. This 420,000-subnode model is input for an AWEsim-based tool to calculate the skew across GCLK. The AWEsim-based tool speeds up the RC network analysis by the RC reduction algorithm using asymptotic waveform approximation, which reduces the high-order RC network to a lower-order network. Figure 7-9(a) shows the simulation of the clock skew across the chip using the tool, and Figure 7-9(b) presents the measurement result. The simulation results in a total skew of 72ps at 100oC and 1.8V assuming the worst-case conditional loading. A detailed RC network is built for each local clock tree starting from either a major clock or GCLK in the Alpha microprocessor chip [98]. Gate capacitance models are placed through the network using the appropriate terminal biasing assumptions. Two SPICE simulations are run for each node, one using minimum effective capacitance assumptions and the other using maximum effective capacitance assumptions. Minimum and maximum delays are specified individually per device load and applied to the timing-based methodology CAD tools for timing path analysis. In all, more than 60,000 local clock nodes are simulated with SPICE in this manner using an aggregate 19 million simulated elements and 14 million simulation nodes. Figure 7-10 shows a histogram from all these simulations of the variation between minimum and maximum delays per local clock. Some variations are as high as 280ps.

Table 7-5. Clock Output Commands [112].

Command	Syntax	Description	Default Value
CLOCK_ ASCII_ OUTPUT_ FILE	CLOCK_ASCII_ OUTPUT_FILE: < file >	ASCII file name of detailed timing simulation results at each stage of the clock network.	None
CLOCK_ SDF_ OUTPUT_ FILE	CLOCK_SDF_ OUTPUT_ FILE: < file >	SDF file generated for the clock network.	None.
CLOCK_ HIGH_ THRESHOLD	CLOCK_HIGH_ THRESHOLD: < Float between 0 and 1>	Percentage of (V_{dd}-V_{ss}) for the end point in rising time measurement.	0.9
CLOCK_ LOW_ THRESHOLD	CLOCK_LOW_ THRESHOLD: < Float between 0 and 1>	Percentage of (V_{dd}-V_{ss}) for the starting point in rising time measurement.	0.1
CLOCK_ SIMULATION _STEP	CLOCK_ SIMULATION_ STEP: < Float >	Time step used in clock simulator transient analysis for accuracy control.	Simulation time divided by 1000
CLOCK_ SDF_ REMOVE_ CELLS	CLOCK_SDF_ REMOVE_ CELLS: <YES | NO>	YES: remove the intermediate clock buffers from clock SDF output file, since the logic netlist may not have those buffers, which are inserted in the physical design stage.	No
CLOCK_ SKEW_ OUTPUT_ FILE	CLOCK_SKEW_ OUTPUT_FILE: <file>	A clock skew summary file from the clock network simulation.	None

7.4 IR Drop Effects

Another issue adding to the complexity of the clock design is the IR drop of the power network. The clock distribution circuit itself is one of the major causes of the IR drop. Clock buffers are designed to switch at high frequencies and to drive large RC loads. The switching of clock buffers is a major source of power dissipation in high-speed chips and leads to the IR drop. The cause of the IR drop is due to the resistance in the power distribution system as shown in Figure 7-11. With the narrow metal lines used in the deep submicron chips, resistance is a major factor in all interconnects including the power distribution system. When large buffers in the clock circuit begin to draw current, the current flowing from the supply to the drivers encounters this resistance and the voltage begins to drop. Therefore, the last inverter, as shown in Figure 7-11, experiences a lower supply voltage than the first inverter. The fluctuation of the supply voltage causes variations in the delay, skew, and slew rates, and results in clock signal jitter.

```
(DELAYFILE
(SDFVERSION "3.0" )
(DESIGN "toprt" )
(DATE "Mon Apr 24 17:56:212000")
(VENDOR "Avant! Corporation" )
(PROGRAM "Star-RCXT" )
(VERSION "2000.4.0.0")
(DIVIDER I)
(TIMESCALE 1ns
(CELL
(CELLTYPE "toprt")
(INSTANCE )
(DELAY
(ABSOLUTE
(INTERCONNECT ctbuflx reg\lq\[0\]lck
(0.441:0.441:0.441)
(10.554:10.554:10.554))
(INTERCONNECT ctbuflx reg\lq\[1\]lck
(0.441:0.441:0.441)
(10.554:10.554:10.554»
(INTERCONNECT ctbuflx reg\lq\[2\]lck
(0.441:0.441:0.441)
(10.554:10.554:10.554))
(INTERCONNECT ctbuflx reg\lq\[3\]lck
(0.441:0.441:0.441)
(10.554:10.554:10.554))
) ) )
(CELL
(CELLTYPE "ctbuf_12" )
(INSTANCE ctbuf)
(DELAY
(ABSOLUTE
(IOPATH ax (0.804:0.804:0.804) (0.947:0.947:0.947))
) ) )
)
```

Figure 7-7. SDF Output File from Clock Analysis [112].

```
* SKEW SUMMARY
* Maximum Skew (Rise:Fall) : 2.0036e-10:1.96001e-10
Minimum Rise Delay: 6.9868e-10 at the following sync
points:
1. clk ( clk )
clkIINV211!O/A ( clk )
clk!INV212!1/A ( clkl1!O )
clk!INV2!311/A ( clkl211 )
clk!INV2!4!3/A ( clk!311 )
reg[2]/ck ( clk!4!3 )
Maximum Rise Delay: 8.9904e-10 at the following sync
pins:
1. clk ( clk )
clk!INV2!1!O/A ( clk )
clkIINV2!2!O/A ( clk!1!O )
clk!INV213!O/A ( clk!2!O )
clk!INV2!414/A ( clk!3!O )
reg[3]/ck ( clk!414 )
Minimum Fall Delay: 1.1035e-08 at the following sync
pins:
1. clk ( clk )
clk!INV21110/A ( clk )
clk1INV2!210/A ( clk!1!O )
clk!INV213!2/A ( clkl2!O )
clk!INV2!416/A ( clk!3!2 )
reg[1]/ck ( clk!416 )
Maximum Fall Delay: 1.1231e-08 at the following sync
pins:
1. clk ( clk )
clk!INV2!1!O/A ( clk )
clk1INV2!211/A ( clkl110 )
clk1INV213!3/A ( clkl2!1 )
clk1INV214!7/A ( clkl3!3 )
reg[O]/ck ( clk!4!7 )
```

Figure 7-8. Clock Skew Summary File [112].

Figure 7-12(a) shows a small clock distribution tree with five stages of clock buffers connected to the Vdd power grid [107]. The clock tree consists of 23 inverters connected together by distributed RC interconnects. The power grid consists of a mesh of 10,000 nodes, 20,200 resistors, and 10,000 capacitors lumped to the ground [107]. Two Vdd power supplies are connected to the power grid at nodes NVdd1 and NVdd2. The asymmetric connection of the voltage sources ensures a variable dynamic IR drop in the power grid. The clock signal is observed at nodes N1 to N8 in the fourth stage.

Figure 7-12(b) shows the simulation results using SPICE for the circuit. A clock input ramp of 0.1ns rise time is applied at node CLK. The output waveforms at four stages are shown, along with the IR drop in the power grid. The highest IR drop value of 0.063V, or 3.6%, is observed at the power grid tap point of inverter G12. The simulation result shows the dynamic behaviour of the power grid, with the maximum drop associated with the switching of stage 4 and stage 5. The effect of the IR drop can be significant on the clock delays. A 5% drop in the supply voltage can cause up to a 15% increase in the clock delays [107]. In addition, the IR drop is usually added non-uniformly to the voltage dips in the full chip, resulting in variable clock delays for the increased clock skew.

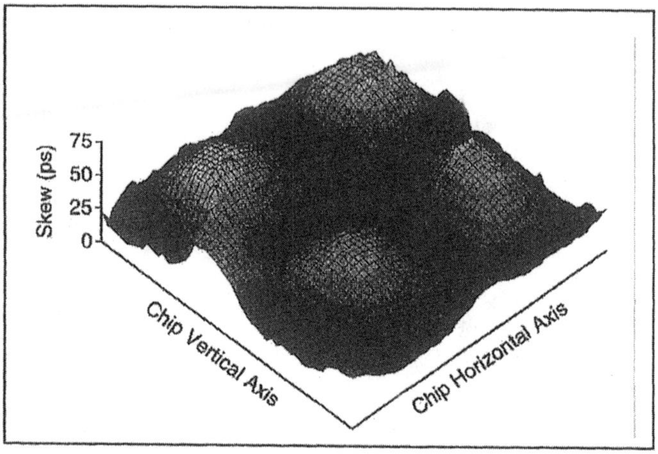

(a)

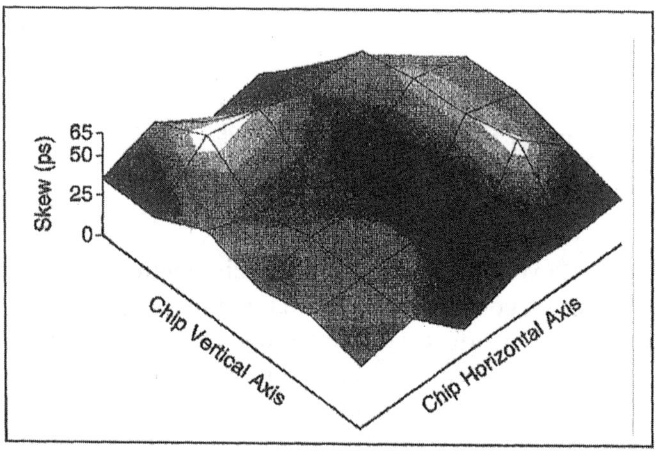

(b)

Figure 7-9. GCLK Skew of Alpha Microprocessor Chip [98].

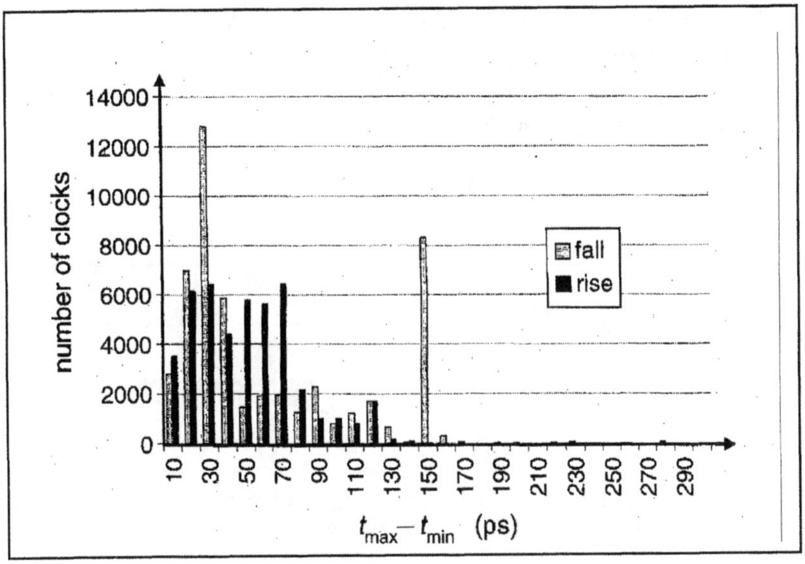

Figure 7-10. Local Clock Skew for Alpha Microprocessor [98].

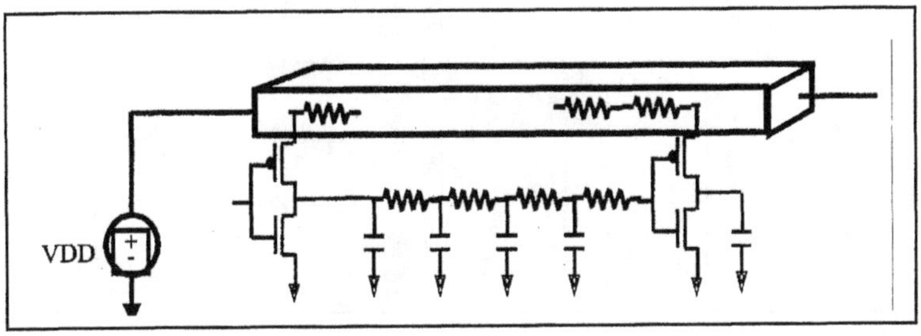

Figure 7-11. IR Drop on Buffer Delays [107].

7.5 Summary

The local clock skew or inaccuracy is usually incorporated into the timing analysis for the setup and hold time results in critical timing paths [62]. With the continually scaled down skew budget in very high frequency design, such as GHz microprocessors, the clock skew and jitter results are tabulated based on the common node in the clock tree for two sequential elements in the circuit timing analysis [58]. If two sequential elements are in close proximity and share most of the clock distribution tree, they will have smaller clock skew requirement between them. Therefore, various clock skew budgets can be used in the timing analysis of critical timing paths. The full chip clock network simulation becomes critical for the above timing analysis methodology. In recent years, EDA companies are providing CAD tools for the clock network extraction and simulation in the full chip scale [107,112].

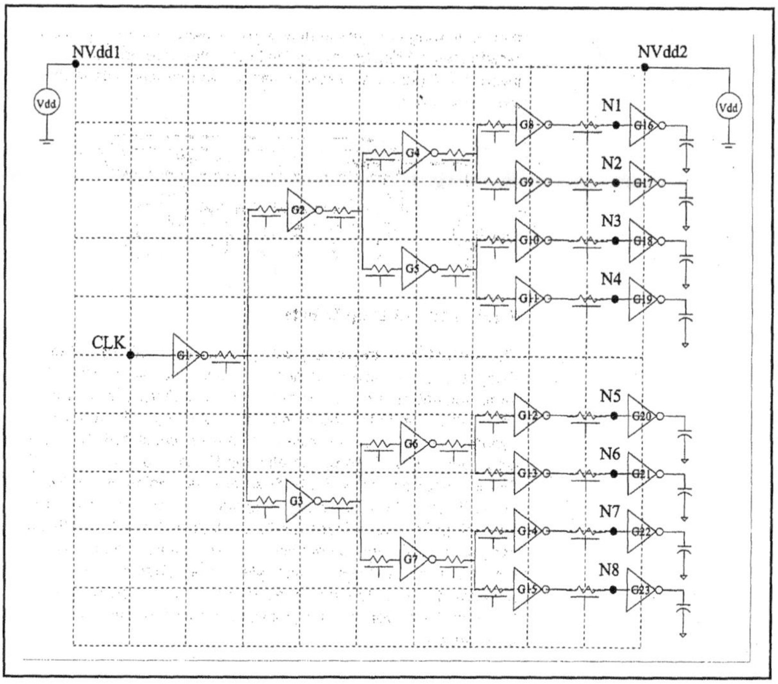

(a)

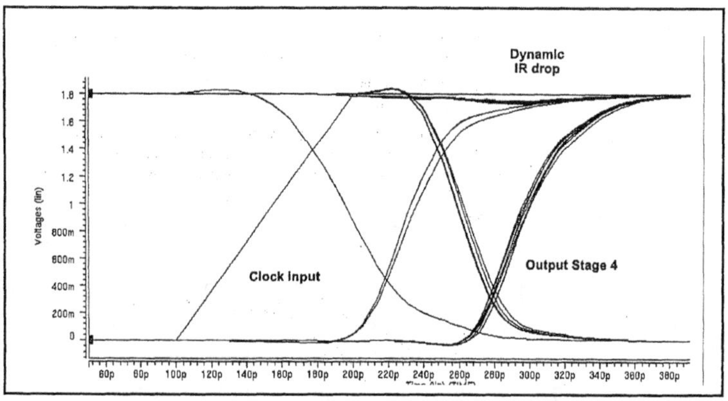

(b)

Figure 7-12. IR Drop Effect Simulation [107].

Chapter 8 Low-Voltage Swing Clock Distribution

The clock network can consume up to 50% microprocessor power based on the experimental data shown in Figure 1-2. The clock gating technique is often used to reduce the wasted power in the clock distribution. This technique adds a stop clock signal for the clock buffer to gate the clock when the clock is not needed for a portion of the clock distribution network. This chapter describes a clock distribution system with low voltage swing clock signals [34]. While the system has an operating voltage generally used throughout the system, at least one of the global clock signal and the local clock signal is a small swing clock signal that has a voltage swing substantially less than the operating voltage of the system. Section 8.1 presents a 1/2Vdd swing local clock distribution scheme. Section 8.2 presents the low voltage swing global clock distribution scheme. Section 8.3 provides a summary to this chapter.

8.1 1/2Vdd Swing Local Clock Distribution

As shown in Figure 8-1 and 8-2, the small swing compatible local clock generator receives a full swing global clock over the global clock network and generates a plurality of reduced swing local clocks on the local clock networks. In Figure 8-2, local_CLK1 and local_CLK1# are overlapped in the common mode, as are local_CLK2 and local_CLK2#. Vdd and Vss in Figure 8-2 show the voltage levels of the waveforms. Latches and other clock control circuits coupled to the local networks are designed to operate properly with the reduced swing clock signals. Figure 8-3 shows the latch used in 1/2Vdd swing clocks. Four 1/2Vdd swing local clocks are generated to present the two non-overlapping phases for clocked elements (latches or flip-flops). The latch is designed to operate with clocks having a voltage swing of approximately one-half of Vdd, the operating voltage.

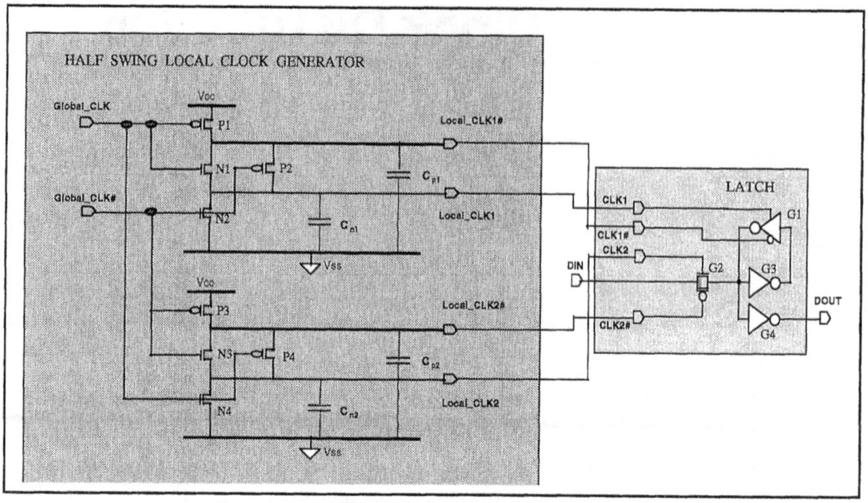

Figure 8-1. Half-Vdd Swing Local Clock Generation and Distribution [34].

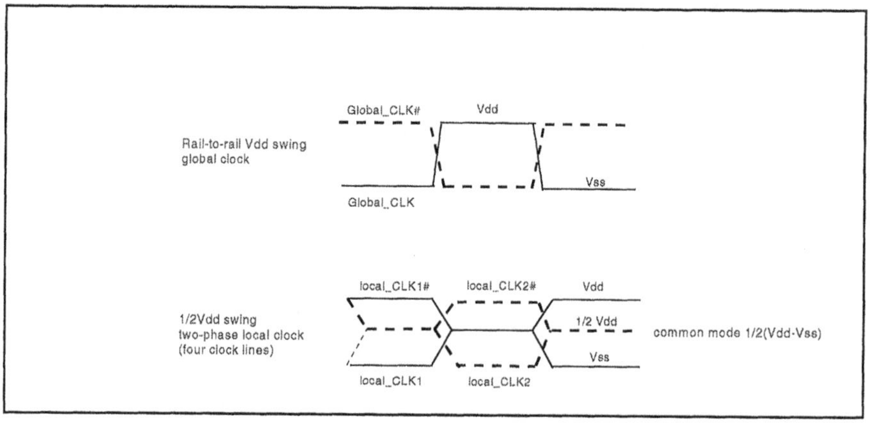

Figure 8-2. Half-Vdd Swing Local Clock Waveforms [34].

Table 8-1 shows the circuit simulation results for the current consumption in both half voltage swing and full voltage swing local clock networks. With the same rise/fall time at clock loads, the 1/2Vdd swing local clock network can save 55% power. To achieve the same rise/fall time, the sizes of transistors are enlarged in the half voltage swing buffer. In the high voltage swing experiment, we use the conventional inverters as the local clock buffers and the latch circuit is driven with two clock lines (CLK1 and CLK1#). The supply voltage is 2.0V in the experiment.

As shown in Table 8-1, the power saving from the half voltage swing local clocks is about 55%, compared to that of using full voltage swing local clocks. The total power includes the clock loads of the latch but not other gates in the latch. Rise time is defined as 20-80%Vmax in both full and low voltage swing cases. In the low voltage swing case, the maximum swing voltage Vmax is the half of Vdd.

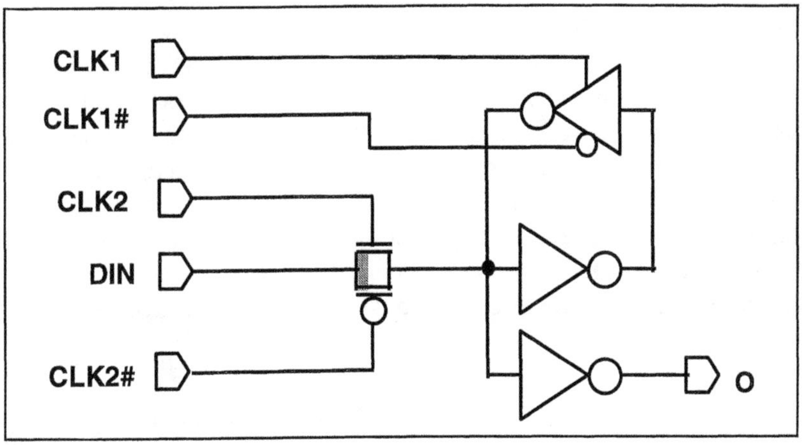

Figure 8-3. Latch Circuit Based on $1/2V_{dd}$ Swing Clocks.

Table 8-1. Comparison between 1/2Vdd Swing and Full-Vdd Swing Local Clock Distributions [34].

	Rise/Fall Time (ns)	*I (Vdd) (mA)*
Full Vdd swing local clock	0.26	650
Half Vdd swing local clock	0.26	300

8.2 Low Voltage Swing Global Clock Distribution

Figure 8-4 shows an example for the low voltage swing global clock distribution network. The global clock generator produces a very low swing global clock signal that is used by the small swing compatible local clock generator to generate full swing local clock signals. These low voltage swing techniques may be used in combination for further power reduction in global and local clock networks. The entire scheme is shown in Figure 8-5. The differential driver generates the low voltage swing clock and this clock is transferred in a global clock distribution network. The differential comparators at the ends covert the low voltage swing clock back to the full voltage swing. The low voltage swing can be as low as 2-3 times less than the normal supply voltage, depending on the noise margin allowed, thus saving significant amount of power in the global clock distribution. There is very little performance penalty for the local clocked elements in using the low voltage swing clock in the global clock distribution, since the local clock signal has been converted back to the full voltage swing.

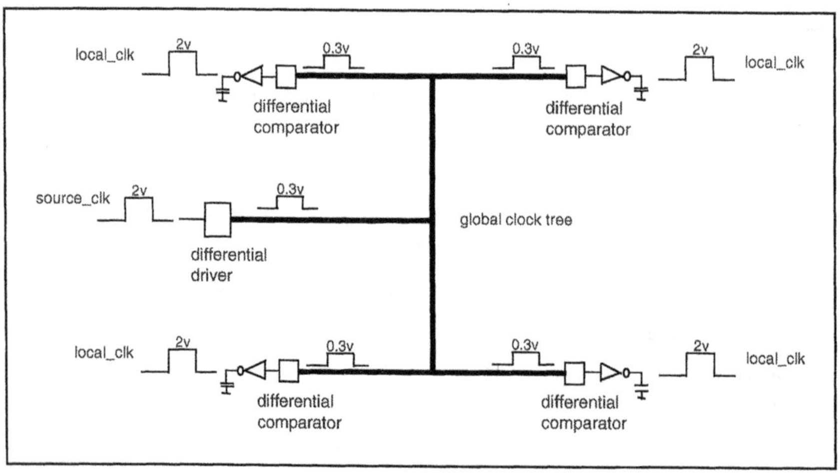

Figure 8-4. Global Clock Network Low Voltage Swings.

One design of the global differential clock driver is shown in Figure 8-6. In this circuit, a PMOS transistor and an NMOS transistor are connected in series between the voltage supply (Vdd) and a bias node. The gate of the PMOS transistor is coupled to its drain, and the gate of the

N channel transistor is coupled to receive an input clock signal CLK. Another PMOS transistor and an NMOS transistor are also connected in series between Vdd and the bias node. The gate of the PMOS transistor is connected to its source, and the gate of the NMOS transistor is coupled to receive an input CLK# signal. A bias transistor couples the bias node to the ground voltage supply line. A small shunt transistor couples the sources of the NMOS transistors and limits the swing of the output voltage of the clock driver. The global clock signal generated at the drain of the PMOS transistor is transferred through a global clock line to the common mode voltage supply and the global clock network. The global CLKOUT# signal generated at the drain of the PMOS transistor is transferred to the common mode voltage supply by a global CLK# line.

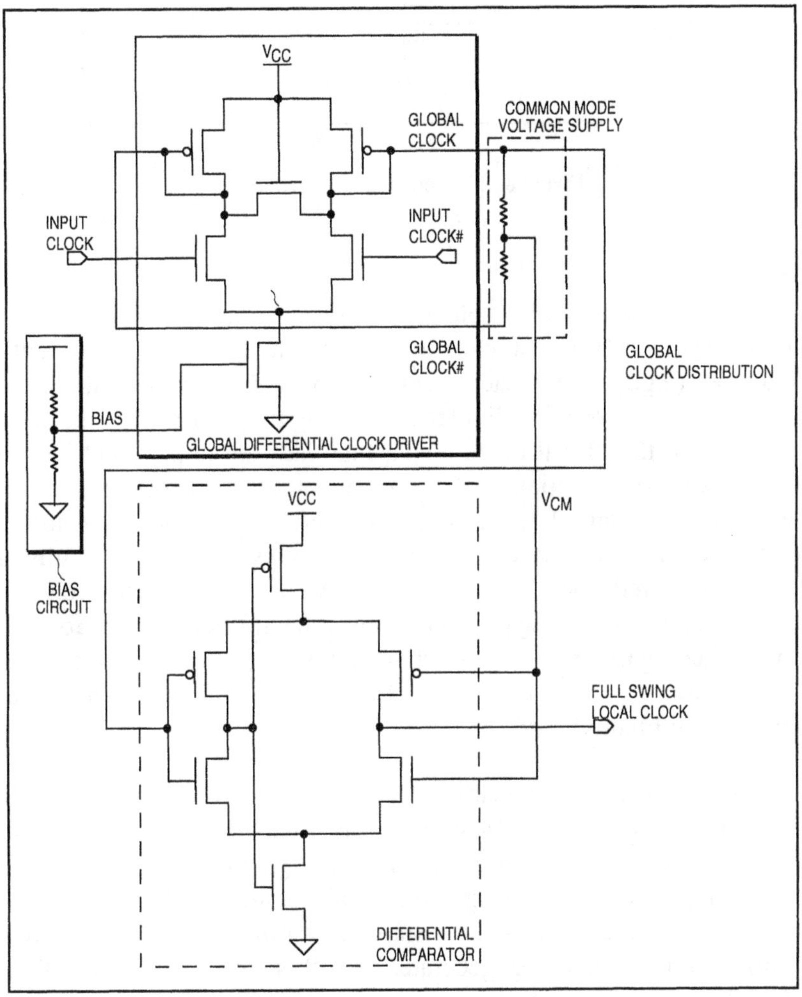

Figure 8-5. Differential Global Clock Distribution Scheme in Low Voltage Swing [34].

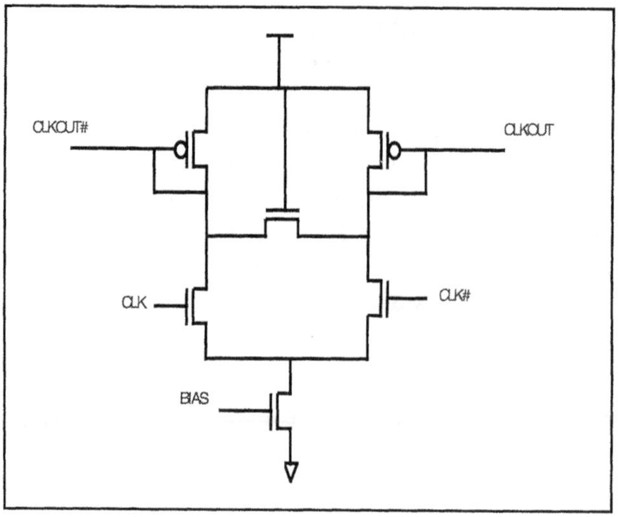

Figure 8-6. Differential Clock Driver.

The common mode voltage supply generates a single common mode voltage (VCM) that can be used by a number of differential comparators (e.g., one for each local clock tree supported by the global clock network). A resistive divider comprising two resistors connected in series between the global clock signal line and the global clock# signal line provides the essentially DC common mode voltage signal on the common mode voltage line. As previously mentioned, the DC nature of this signal obviates the need for balanced or other specialized routing as would be required for an additional clock signal. Alternatively, the common mode voltage supply may simply be formed locally for each differential comparator using a resistive divider to divide down the system voltage source (Vdd). The ideas for VCM generation circuit are summarized in Figure 8-7.

The differential comparator, as shown in Figure 8-8, is exemplary of a variety of traditional comparators, which may be used to recover the full swing clock signal from the small swing global clock signal. This circuit is a kind of self-biasing differential comparator [45]. From the small signal global clock signal transmitted over the global clock network, the differential comparator generates a full swing local clock signal CLKOUT.

Figure 8-9 shows the clock waveforms at various stages of the global clock network using the differential buffers. V(clk#) is the input clock# signal to the differential clock driver (the clock signal, not shown here, is at a 180-degree phase reverse to clock#). V(begin_clk) is the low voltage swing clock after the differential clock driver. V(diff-in) and V(diff-inb) are two complementary clock inputs to the differential comparator. V(diff-out) is the full voltage swing clock after the differential comparator. V(last-out) is the full voltage swing clock after stages of local clock buffers followed by the differential comparator. V(load_clkf) is the clock waveform at the clock load in local clock network. The end clock signal V(load_clkf) is in full voltage swing; therefore there is a little impact on the performance of latches or other clocked elements.

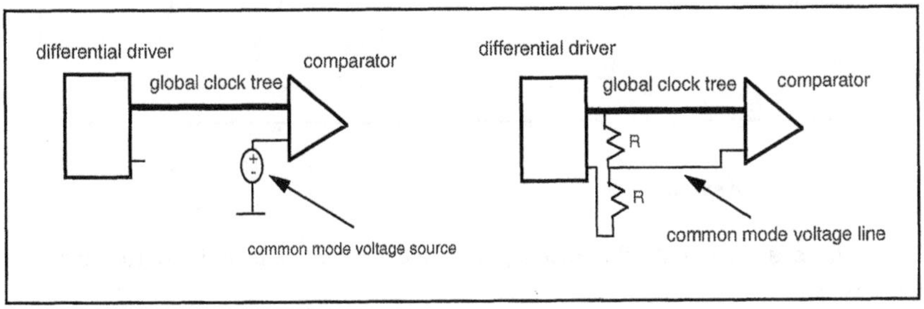

(a) Local Generation (b) Global Generation

Figure 8-7. Reference Clock Generation for Differential Comparator.

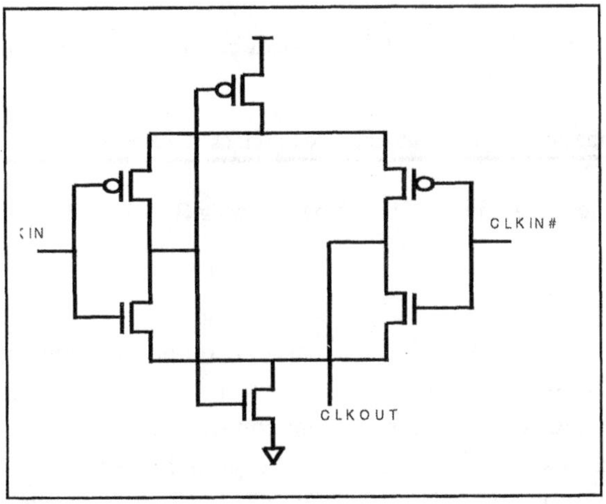

Figure 8-8. Differential Self-Biased Comparator [45].

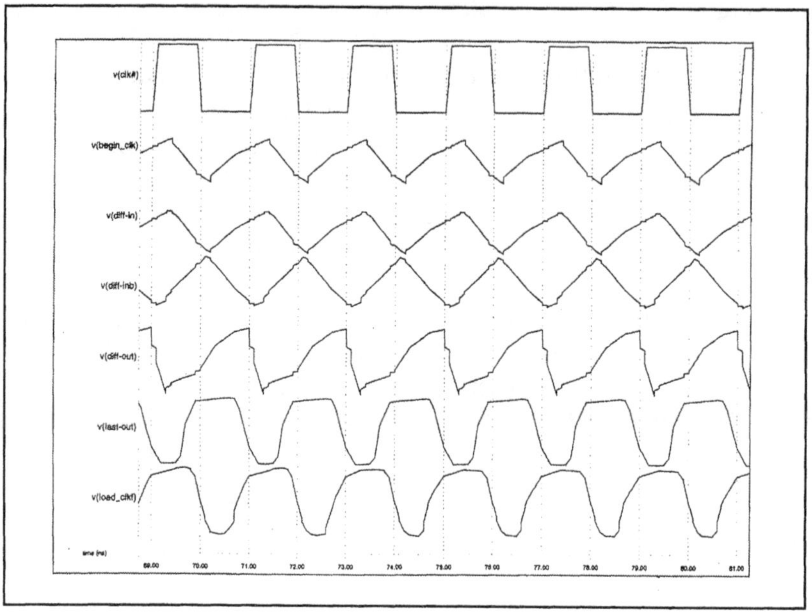

Notations of the waveforms:

V(clk#): the clk# input to the differential global clock driver (clk input is not shown but with reverse phase of clk#).

V(begin_clk): the immediate output of the differential global clock driver.

V(diff-in), V(diff-inb): two clocks as the inputs of the differential comparator.

V(diff-out): the immediate output of the differential comparator.

V(last-out): the immediate output of the local clock buffers after the differential comparator.

V(load_dclkf): the clock signal at the input to the clocked element.

Figure 8-9. Clock Waveforms of Low-Voltage Swing Global Clock Distribution [34].

Table 8-2 shows the current consumption comparison for the clock network with the low voltage swing global clock and full-Vdd swing global clock. The current consumption includes the entire clock distribution network, which contains the global clock drivers, the global clock distribution tree, the local clock buffers, and the local clock network. Wire widths of the global clock tree are scaled to offset the impact on the power reduction due to the wire capacitance reduction. The low voltage swing global clock scheme, including the four sizes of global clock trunks, saves 57% power on average. Although analog-type differential converters consume a bit more power, due to the cross-bar or short-circuit currents, than the conventional digital buffers containing the inverters only, the experimental results show the overall power reduction using the low voltage swing global clock is about 57% in the microprocessor case. The low-voltage swing global clock reduces the power consumption for the heavy capacitance load in the global clock routing tree. The low-voltage swing global clock also improves the rise/fall time. As shown in Table 8-2, the rise time Tr at the inputs of the local clock buffers is reduced by 55% by using the low voltage swing global clock. The rise time improvement is mainly due to the RC delay reduction using the low voltage swing. One more interesting observation from Table 8-2 is that the rise time in the low voltage swing scheme is almost not affected by the trunk width reduction or wire RC increase. Rise time T_r is defined as 20-80%Vmax in both full and low voltage swing cases.

Table 8-2. Experiments of Low Voltage Swing Global Clock Distribution [34].

I(Vdd) (mA)	Original trunk width	1/2 Width	1/4Width	1/16 Width
Low Swing	91.5	86.9	82.3	77.9
Full Swing	265.6	198.5	189.7	182.5
T_r (ns)	Original trunk width	1/2 Width	1/4 Width	1/16 Width
Low Swing	0.21	0.20	0.19	0.18
Full Swing	0.27	0.47	0.58	0.71

8.3 Summary

The low voltage swing circuits, presented in this chapter, include the differential circuits for global clock trees, and the charge-sharing local clock regenerators for the clocked latches based on two non-overlapping 1/2Vdd swing local clocks. In order to achieve the power saving without the performance degradation of clocked elements, the global clock network becomes the preferred candidate for using the low voltage swing technique. The differential circuits also provide the high-speed improvements for global clock distribution. Generally, the use of differential signals requires the transmission of a signal and its complement. In the area of clock distribution, the layout area overhead of two balanced clock trees may be quite substantial. Although the differential low voltage swing global clock circuit reduces the power by 57% and improves the clock rise time by 55%, it introduces 10% more clock skew [34]. The clock skew is due to more transistors and more sensitivity to the process variations in the differential circuits. Special design considerations are necessary for the clocked elements and the local clock buffers. Noise is always an issue in the low voltage swing circuit operations, although the differential clocks in the global clock distribution scheme help to cancel out the common-mode noise. Careful shielding of the clock lines is required in any case to prevent coupling from adjacent lines.

Chapter 9 Routing Clock On Package

Traditionally, the global clock distribution tree or network is routed on the top metals of the chip. The metal thickness and width in the highest metal layer are larger than those in the lower layers. Furthermore, the wires comprising the package layer are even wider and thicker, with usually a 1-2 order larger line scale than the on-chip interconnects. The interconnect resistance on package is 2-4 order less than the on-chip metal resistance. Based on flip chip technology and area I/Os, the global clock tree or network can be routed on the package. The layout design of the chip becomes easier by separating the global clock network from the rest. The clock trees on the package layers still have few obstacles to overcome before it is applied to the real chips, which will be studied in this chapter. Section 9.1 presents an overview. Section 9.2 discusses the ESD design. Section 9.3 shows the noise considerations. Section 9.4 shows the experimental results on a microprocessor test chip. Section 9.5 provides a summary to this chapter.

9.1 Scheme Overview

As was discussed above, it is desirable to keep the wire resistance and capacitance of a clock tree to a minimum in order to optimise power, skew, and the transition time parameters. This may be accomplished by the package clock distribution as illustrated in Figure 9-1(a) [15,18,122]. The global part of the clock tree/network is routed on the package layer and the local part of clock trees are on chip. Typically the package layer comprises copper, which has lower resistance than the integrated circuit interconnect. The package layer interconnects are typically wider and thicker than those implemented on the integrated circuits. Thus the resistance is reduced for the global clock tree, which is usually 2-3 orders less than that of the integrated circuits. Another advantage of using the package layer is that it comprises a thick layer of insulator, as shown in Figure 9-1. Thus the capacitance of the global clock tree is decreased, decreasing power consumption and rise time.

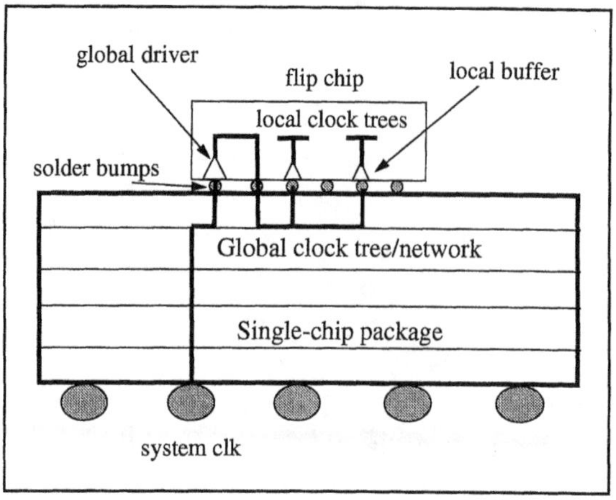

(a) Single Chip

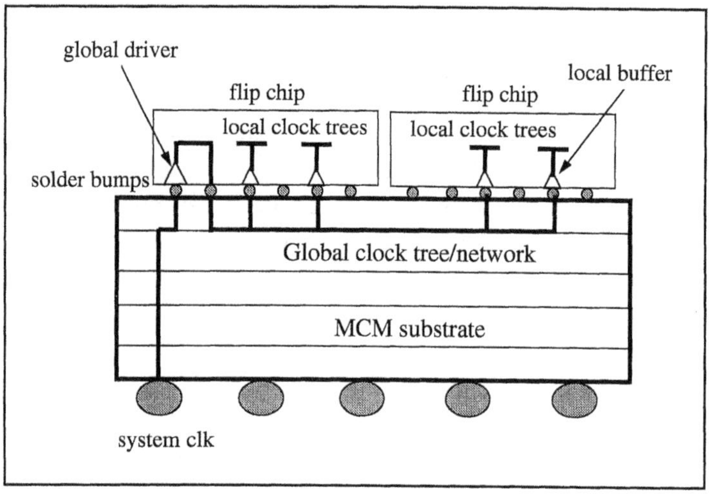

(b) Multi-Chip Module (MCM).

Figure 9-1. Chip and Package Clock Network Co-design [15,18].

As shown in Figure 9-1(a), the global clock tree on package and on-chip global driver and local clock trees are connected by area or array solder bumpers provided by the flip chip technology. The flip chip technology provides area I/Os that are distributed over the entire chip surface rather than being confined to the periphery. The package and the integrated circuit are coupled by the area contacts or solder bumpers.

Compared with the wire bonding and TAB, the flip chip technology has the highest I/O density, smallest chip size, and lowest inductance. As illustrated in Figure 9-1(a), the clock source resides on the integrated circuit and is routed to a solder bumper. The local clock trees and local clock buffers are routed within the integrated circuits and connected to solder bumpers and then to the on-package global clock tree.

The clock source (global driver) represents a circuit for generating an on-chip clock signal, such as a phase-locked loop (PLL). The clock source receives a system clock signal from an I/O port of the package, coupled to an external clock source such as an oscillator. Although Figure 9-1(a) illustrates a ball grid array package (BGA), it should be noted that other package types may be used for this technique, such as pin grid array (PGA), dual inline package (DIP), or plastic quad flat pack (PQFP). This design technique can also be used for the clock distribution in multichip modules (MCM) environment. The local clock trees in multiple chips are connected to the global clock tree routed on the MCM substrate, as shown in Figure 9-1(b).

9.2 ESD Design

When the global clock network is placed on the package layers, the local clock buffers are required for the protection with the ESD (electronic-static diodes). Figure 9-2 shows the ESD circuit structure, which can be used and simulated in our experiments. The primary sources of clock skew are as follows: the variations in transistor characteristics, the variations in loads, and the variations in the interconnect characteristics. For example, the channel length variation across the die will lead to a clock skew. Other factors affecting the transistor are cross-die power supply voltage variation and temperature variation. For interconnects, variations in the thickness is a primary cause. The combined effects of metal resistance variation and thickness variation is estimated to contribute to approximately 25% difference in the interconnect performance. Lastly, differences in local clock loading also contribute to clock skew. Table 9-1 summarizes the on-chip variation used in our experiments. As seen from the table, the effects of in-die Vcc and temperature variations are lumped in with the variation in the channel length (drawn). A total of +/- 0.03um is used in simulations.

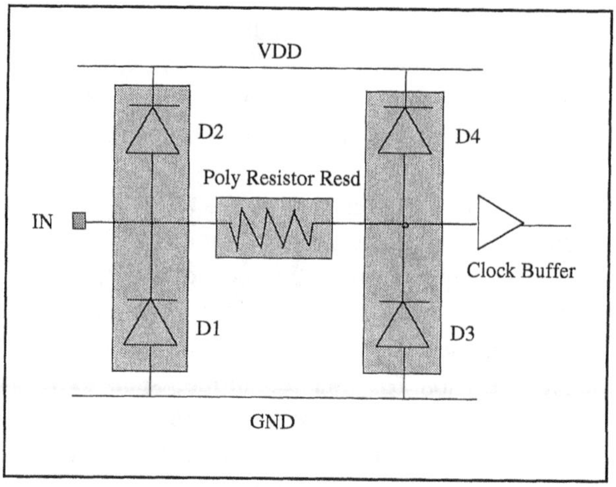

Figure 9-2. ESD Circuit for Local Clock Buffers [18].

Table 9-1. In-Die Transistor, Interconnect and Load Variations [18].

	Variation	**Comment**
Channel length	0.025μm	nominal = 0.4um drawn
Vcc	+/- 50mV	reflected as +/-0.0029um on L drawn
Temperature	+/- 20C	reflected as +/-0.0021um on L drawn
Tox	0%	
Metal sheet resistance	+/- 4%	
Wire thickness	+/- 8%	
Region-to-region load	+/- 20%	nominal load is 4nF/chip

The simulation model for the entire clock tree with the package layers is illustrated in Figure 9-3. It consists of two paths: slow and fast. The wires on the package are modelled with distributed RLC modules. The same variation on transistors and on-chip interconnects are assumed in Table 9-2. The size of the global driver is reduced because the capacitance of the package clock tree is lower than that of the on-chip clock trees. Table 9-2 shows the skew (rising edge at the load) for five Wnglobal sizes (global driver sizes). A final value of 430um is chosen for the global driver size.

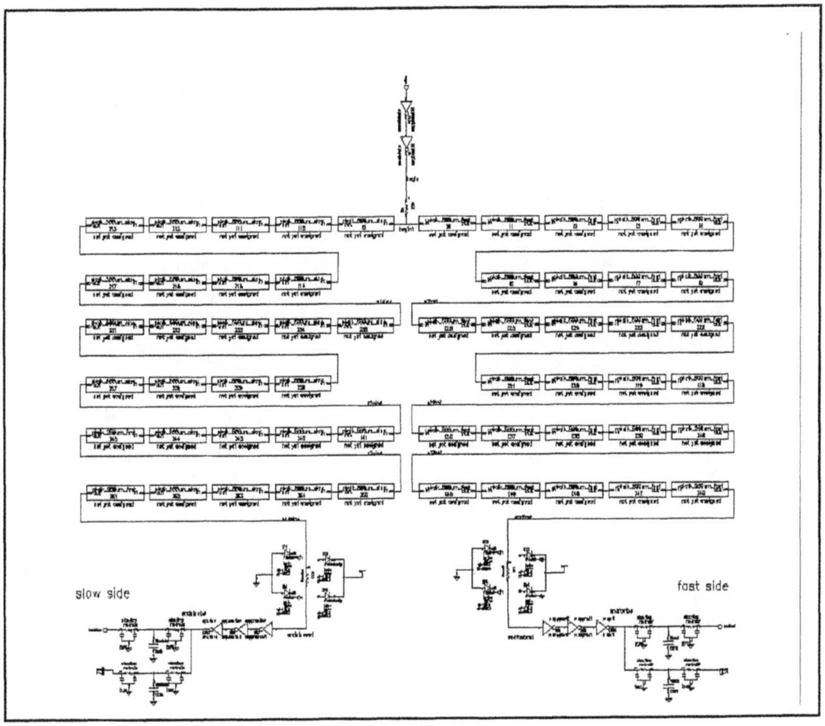

Figure 9-3. Package Clock Routing Simulation Schematic (Stage Ratio ~ 3.5, P/N ratio = 2.8) [18].

Table 9-2. Skew (rising edge) vs. Global Driver Size (Wnglobal) (Rs = 0 ohm, no ESD) [18].

Wnglobal (μm)	200	300	375	430	500
Skew (rising edge) at Load (ps)	235.4	172.1	153.9	146	138.2

Figure 9-4 shows the relationship between the ESD resistance (Resd) and the skew (rising edge at the load). Resd variation also contributes to the total skew. Two cases are simulated with a series termination resistor (Rs = 12.5 ohms) and without a source series termination (Rs = 0). When Rs = 0, the skew is decreased by about 80ps compared with Rs = 12.5 ohms. The ESD circuits provide the parallel termination (i.e., Rs = 0) of the global clock tree. The skews at the loads (rising edge) are 293ps and 216ps, respectively. The data in Figure 9-4 suggests that minimal clock skew is obtained with low Resd (about 5 ohms) without the need of the source termination resistor. The clock skew will increase with the larger Resd values because of the Resd process variation.

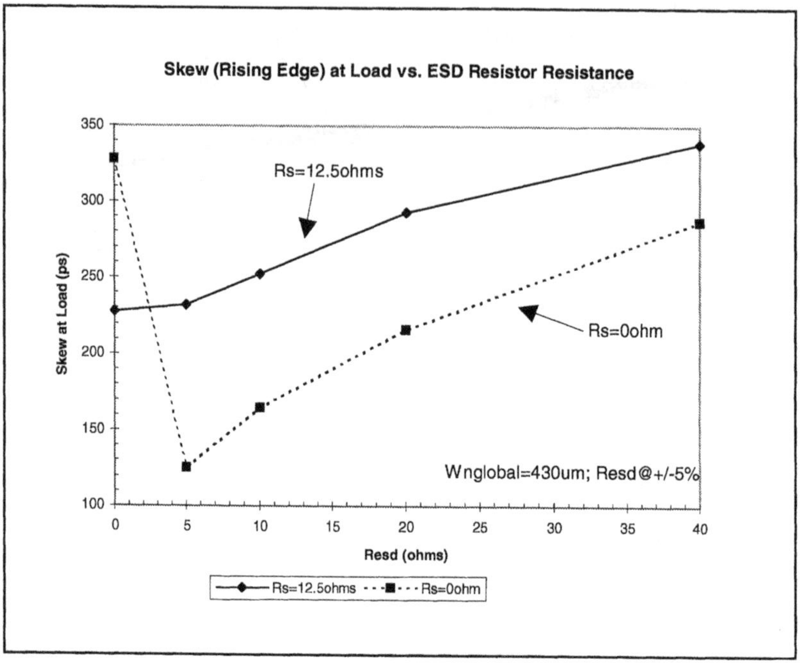

Figure 9-4. Skew (Rising Edge) vs. ESD Resistor Value (Resd variation @+/-5%) [18].

Since the global clock tree and local clock trees are self-connected between chip and package, the ESD is not required for the human-body touch protection. ESD protection is needed in the package assembly phase when attaching the chip to the package. Therefore, if one can control the manufacturing environment well, small size ESD diodes and low ESD resistance may be enough. Although the pass voltage of ESD reliability for human body test is 4000V, the device model test is about 1000V. Different alternatives for choosing the diode sizes and ESD resistances are also presented in Table 9-3. With small size diodes and 5 ohms series resistance, the ESD circuit causes the smallest clock skew with a small device area. Figure 9-2 shows the schematic for a standard cell implementing the local clock buffer ESD. The ESD structure is broken down into three parts, allowing the designer to take full advantage of the flexibility in the ESD guidelines. The D3 and D4 diodes must stay close to the gates they are protecting. D3 and D4 may have to be moved independently of D1 and D2 diodes. This is particularly useful when the placement of the local clock buffer is far away from the solder bumper that usually has been fixed in an array on the chip surface. The ESD series resistor is also a subcell, which can be optimised between the solder bumper and local clock buffer to minimize the clock skew. By "snaking" the ESD diodes and polysilicon resistor into the spare layout area, the die size can be reduced with a small penalty in ESD area.

Table 9-3. ESD Circuit Choices [18].

Resistance	5 ohms	5 ohms	100 ohms
Diodes Needed	D1 D2	D1 D2 D3 D4	D1 D2 D3 D4
Diode Sizes	Small	D1 D2 large, D3 D4 small	Large
Risk	Medium	Low	Proven no risk
Skew and delay	Good	Good	Bad

9.3 Transmission Line Noise on Package

The simulation is also done without the ESD circuits in the local buffers. Figure 9-5 shows the simulated waveforms. The skew at the load is small and in part due to the local clock buffer. However, the signals at the inputs to the local buffers exhibit overshooting and ringing waveforms because of the transmission line reflection noise of the package clock tree. Signal quality is improved with the addition of source series termination, as shown in Figure 9-6. The skew at the loads, however, increases with the series termination resistance (Rs). Figure 9-7 shows the relationship between the series termination resistance and the skew at the load. Simulations show that there is little difference in the quality of the waveforms for Rs between 10ohms and 20ohms. Since the skew will be increased with Rs, an Rs with 10-13ohms is selected based on the simulation results.

The general method to determine Rs is as follows. Rs is sampled from Rmin to Rmax in increments of Rinc. Here, Rmin is the smallest possible series termination resistance (usually 0), Rmax the largest possible series termination resistance, Rinc the increment of resistance sampling, and Rmax = Rmin + n * Rinc. Suppose the maximal overshoot and maximal undershoot are below the required bounds within the range [Rs1, Rs2]. Then Rs1 is the optimal resistance selected for the series termination resistor since it yields the smallest clock skew within [Rs1, Rs2]. Note that the skew is monotonically increasing with Rs, as shown in Figure 9-7.

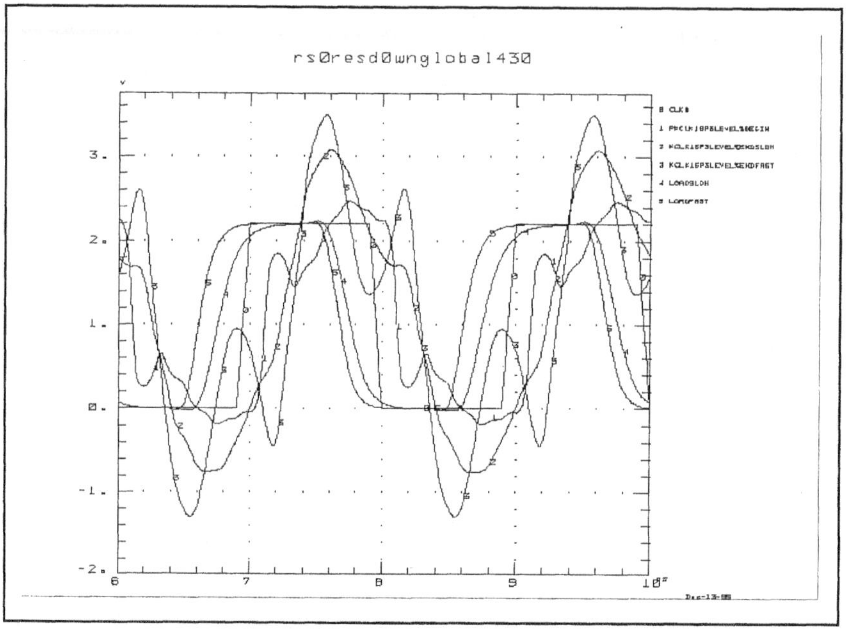

Figure 9-5. Simulated Waveforms for Package Clock Routing (no source termination, no ESD) [18].

9.4 Microprocessor Experimental Results

Table 9-4 shows the simulation results of a microprocessor test chip using the package clock distribution. It shows the simulation results when a series termination resistor (Rs = 12.5ohms) for the package noise in the simulation. As mentioned earlier, the potential benefit of package clock routing is the lower package trace resistance and capacitance. Lower trace resistance and capacitance allows a reduction in the number of local buffers to achieve a lower skew. For example, the three levels of local buffer units contribute 165ps ~ 181ps of the clock skew. Reducing these to two levels can reduce the skew by 40ps ~ 55ps. To accommodate the higher load at the end of the H-tree, the size of the global driver is increased. The global driver size (Wnglobal) is increased from 430um to 1128.8um and also to 2257.6um.

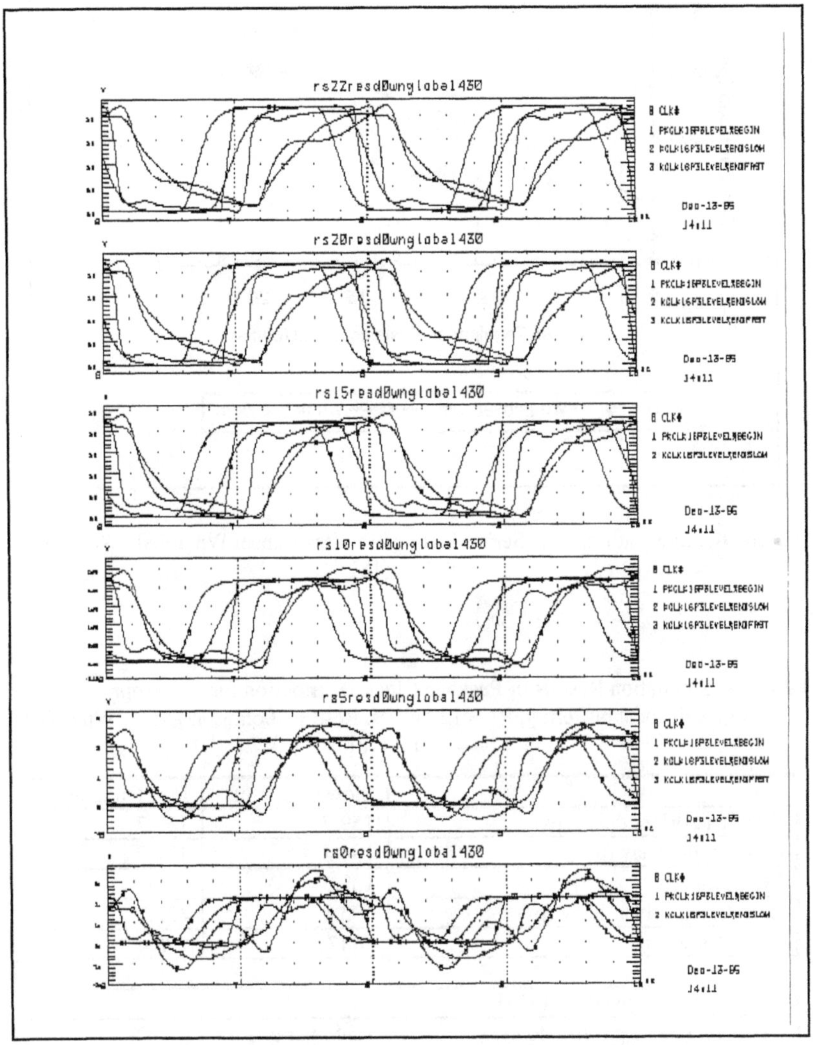

Figure 9-6. Clock Waveforms with Different Source Termination Resistors (Rs : 0, 5, 10, 15, 20, & 22 ohms; no ESD) [18].

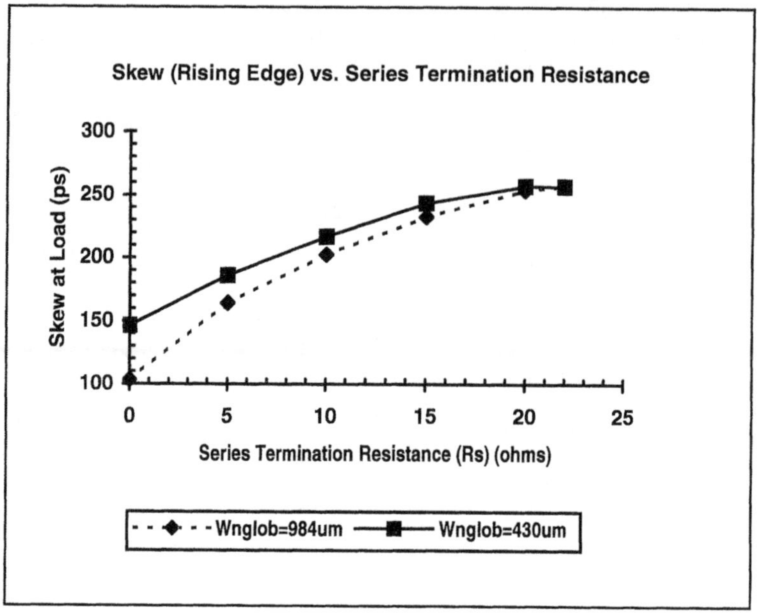

Figure 9-7. Skew at Load vs. Series Termination Resistance: Wnglobal = 984um and 430um [18].

Table 9-4. Simulation Results of Package Clock Distribution for a Microprocessor Test Chip. (3 levels local buffering, Rs = 12.5ohms, Resd = 20ohms @+/-5%, Wnglobal = 430um) [18].

	Slow Quadrant	Fast Quadrant
Front-end driver delay (ps)	159.7	159.7
Clock trunk delay (ps)	184.8	134.9
Resd delay (ps)	139.3	81.3
Local clock driver delay (ps)	603.8	422.7
Total delay (ps)	***1087.5***	***798.4***
Skew (ps) average HL/LH (LH)	289.1 (293)	
Icc (max) (A) (avg.(fast, slow))	1.544 per quadrant	
Icc (rms) (A) (avg.(fast, slow))	0.694 per quadrant	
Icc (avg) (A) (avg.(fast, slow))	0.502 per quadrant	
Itrunk (avg) (avg.(fast, slow)) (A)	0.00983 per quadrant	
Iglobal (avg) (avg.(fast, slow)) (A)	0.1354 per quadrant	

Since the clock network contributes to a good portion of the total chip power consumption, effort is needed to reduce the capacitance of interconnects and loads in the clock network. By using the thicker dielectric insulator, the package layer can provide lower trace capacitance than that on chip. Therefore, routing the clock network on package layers reduces power and lowers trace capacitance. Table 9-5 shows the average power consumption of the microprocessor clock distribution system: (a) the whole chip including all the loads (Total Power), (b) the global clock tree with the local buffer's input gate capacitance (Global Power), and (c) the global clock tree (interconnects) without the local clock buffer's input gate capacitance (Trunk Power). A 10% power saving for the whole clock system, including the capacitance load, is obtained for this microprocessor test chip.

Table 9-5. Clock Power Consumption Comparison (tttt, 2.2V) [18].

	On-chip Clock Routing	**Package Clock Routing**	**Reduction**
Total Power (avg) (W)	19.71	17.67	10%
Global Power (avg) (W)	6.55	4.77	27%
Trunk Power (avg) (W)	1.89	0.35	81%

9.5 Summary

With continually increasing difficulties of clock distribution in high-speed microprocessors and ASICs, the package clock distribution shows very promising advantages. The concurrent design of chip and package provides optimal design of the clock network by making use of the advantages of package layers, which provide 1,000 times less wire resistance and 10 times less wire capacitance than those of interconnects on chip. Therefore, it is more beneficial to route the global clock network on package. The package clock distribution has the additional advantage that it alleviates the on-chip routing congestion. The chip floor planning and layout are made easier by removing the global clock trunks from design considerations. The chip sort test can be done with a global clock tree connected on the probing card. We believe that package clock distribution technology will be useful with continually increasing clock frequencies in microprocessors. There are additional considerations in the design and testing when routing portions of a clock network on-chip and other portions off-chip. Electrostatic discharge (ESD) protection has to be used for local clock buffers. The local buffer gates could be damaged by ESD if they are not properly protected. The global clock tree on package shows overshoot and ringing clock waveforms. This is due to transmission line reflection noises caused by the low impedance (low resistance and low capacitance) and the inductance effect on package layers. Termination techniques are used to reduce transmission line noise. This chapter provides preliminary study results on the above issues for a global clock tree in a microprocessor test chip.

Chapter 10 Balanced Clock Routing Algorithms

The balanced clock routing algorithms have been researched by many people [7-16]. The goal is to develop a CAD tool, which can route the clock trees in the balanced form to minimize the clock skew. We will summarize the CAD algorithms developed by the author in the Ph.D. dissertation [122, 15,29]. Section 10.1 shows an algorithm to construct a planar equal path length clock tree. Section 10.2 discusses the embedding approach to transform an equal path clock tree with Manhattan wires. Section 10.3 shows a skew-bounded clock tree refinement method. Section 10.5 describes wire sizing technique and optimization flow for the clock skew reduction. Section 10.6 provides a summary of this chapter.

10.1 Planar Equal Path Length Clock Routing

Planar Equal Path Length Clock (Steiner) Tree Routing Problem: *Given a source point s and a set of sinks, find a planar clock (Steiner) tree T, with the minimum total wire length, such that the path lengths from s to all sinks are exactly the same.*

The above problem can be proven to be NP-hard [122]. An NP-hard problem is one for which there are no known optimal solutions in the polynomial time complexity related to the increasing problem size. We present an algorithm to construct a planar equal path length tree. But the total wire length of the clock tree may not be minimal. The path length from the source to destination is minimized.

We construct a planar equal path length rectilinear clock tree in two steps. First, we construct a planar equal path length clock tree, with the connections from the source to sinks in straight lines. Then, we use the second step to transform each straight line constructed in the first step into a set of rectilinear (x-y Manhattan) lines. The VLSI routing for layout implementation prefers the x-y Manhattan routing. The second step will maintain the planarity and equal path lengths, and the total wire length is usually reduced from Step 1 to Step 2 because of the sharing of some wire segments in the x-y embedding process.

The algorithm in Step 1 for constructing a planar equal path length clock tree is a single tree's growth method. This algorithm starts from the source, and sinks are added to the tree gradually. The algorithm produces a series of partial trees $\{T_1, T_2, \ldots, T_n\}$, where T_i is the partial tree, in which the first i sinks are connected. T_n is the final Steiner tree with n sinks connected in the clock tree. The details of the algorithm are as follows. We first connect the sink, which has the maximum Manhattan distance to the source. This forms the first partial tree, T_1, as shown in Figure 10-1(b). As the tree grows, all sinks are classified into two types. A sink is called a free sink, if it has not yet been connected in the tree; otherwise it is called a connected sink. At any partial tree T_i $(i < n)$, we select a free sink and connect it to a branch of the tree, by maintaining the planarity and equal path lengths of the tree. This sink then becomes a connected sink. This may split the branch in two, since a Steiner point is inserted on it. Proceeding from Figure 10-1(b) to Figure 10-1(c), sink t_2 is chosen and connected to branch b_1. A new Steiner point s_{21}, inserted on b_1 for t_2, is said to be a balance point, since s_{21} has equal Manhattan distance to sink t_1 and sink t_2. Generally, a balance point s_{jk} exists for a free sink t_j on a branch b_k, if the Manhattan distance from s_{jk} to t_j is equal to the path length from s_{jk} to any connected sink in the tree. Furthermore, the balance point s_{jk} is called a feasible balance point of t_j if a straight line connecting t_j and s_{jk} crosses no other branches. Connecting a sink t_j to its feasible balance point in tree T_i forms a new tree T_{i+1}, in which both planarity and path length balance are maintained.

A free sink t_j usually has several feasible balance points in a tree. We define the feasible balance point, with the minimum Manhattan distance to sink t_j, as the minimal balance point of t_j in T_i. The Manhattan distance between t_j and its minimal balance point is called the minimal balance distance of t_j. Two rules play key roles in this algorithm as follows:

- Min-rule: Always connect a free sink to its minimal balance point.
- Max-rule: At each stage, always select the free sink whose minimal balance distance is the maximum among all remaining free sinks.

The max-rule guides the connection order for free sinks. This rule ensures the planarity of the equal path length Steiner tree without iterations. The min-rule reduces the total wire length of the balanced tree. It has been proven that the max and min rules combined will always produce a planar equal path length clock tree for arbitrary locations of the clock source and clock sinks [15]. The algorithm grows the clock tree by iteratively applying the max-min rules. At a partial tree T_i, a free sink with the maximum minimal-balance-distance among all free sinks is connected to its minimal balance point. This free sink then becomes a connected sink and T_i is expanded to T_{i+1}. The process continues until all sinks are connected. The algorithm of constructing a planar equal path length clock tree is summarized in the following pseudo code.

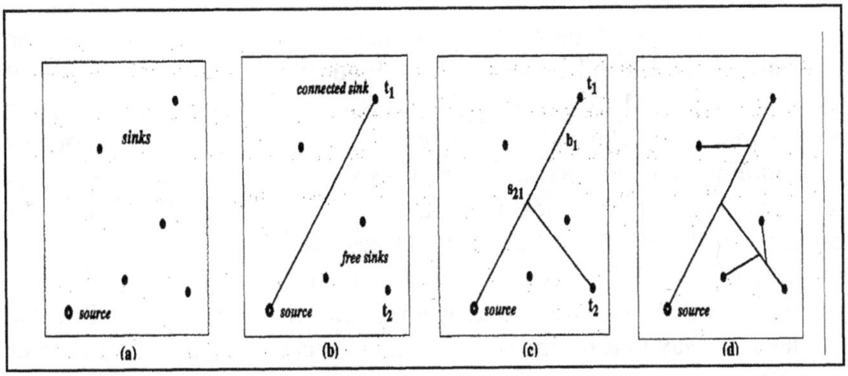

Figure 10-1. Planar Balanced Clock Tree Growing [15].

Planar Equal Path Length Clock Tree Construction Algorithm

Input: a source s, and a set of sinks D;

Output: a planar equal path length clock tree T.

Procedure *ClockTreeConstruction*(s, D, T) {

C_0 = D;

T = ({s}, 0);

CreateBranch(C_0, T);

}

Procedure *CreateBranch*(C, T){

Find t*, b* such that: d(t*,b*)=$\max_j\{\min_i\{d(t_j,b_i),b_i$ is a bounding branch of $t_j\}, t_j$ in C};

Create a branch from t* to its minimal balance point on b* resulting in new T;

```
If (C-{t*} is not empty) {
        Partition C-{t*} into sub-clusters C1, C2, ..., Ck using
        the reachable relation;
        For each i in [1, k]
                CreateBranch(Ci, T);
}
}
```

The details of the hierarchical max-min optimization used in the above pseudo code can be found in [15]. Figure 10-2 shows the clusters of free sinks separated by the growing clock tree. For each cluster we can select a free sink independently, based on the max-min rules, to connect to the minimal balance point. Figure 10-3 shows an example to construct the planar equal path clock tree based on the proposed algorithm as described in the above pseudo code. Overall the time complexity of the tree growing is $O(1*n)$, where l is the number of levels of the tree growing. In the worst case, at each level, branches are connected to the tree in such away that no further partition of clusters occurs. This implies $l = n$. So, the worst-case running time of the algorithm is $O(n^2)$. However, l is usually much smaller than n for large examples. For example, $l = 21$ and $n = 269$ for the Primary1 benchmark example, and $l = 28$ and $n = 603$ for the Primary2 example [15].

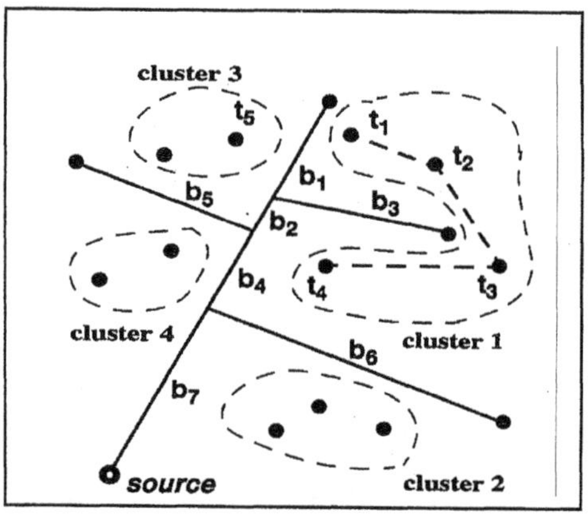

Figure 10-2. Four Clusters for Free Sinks.

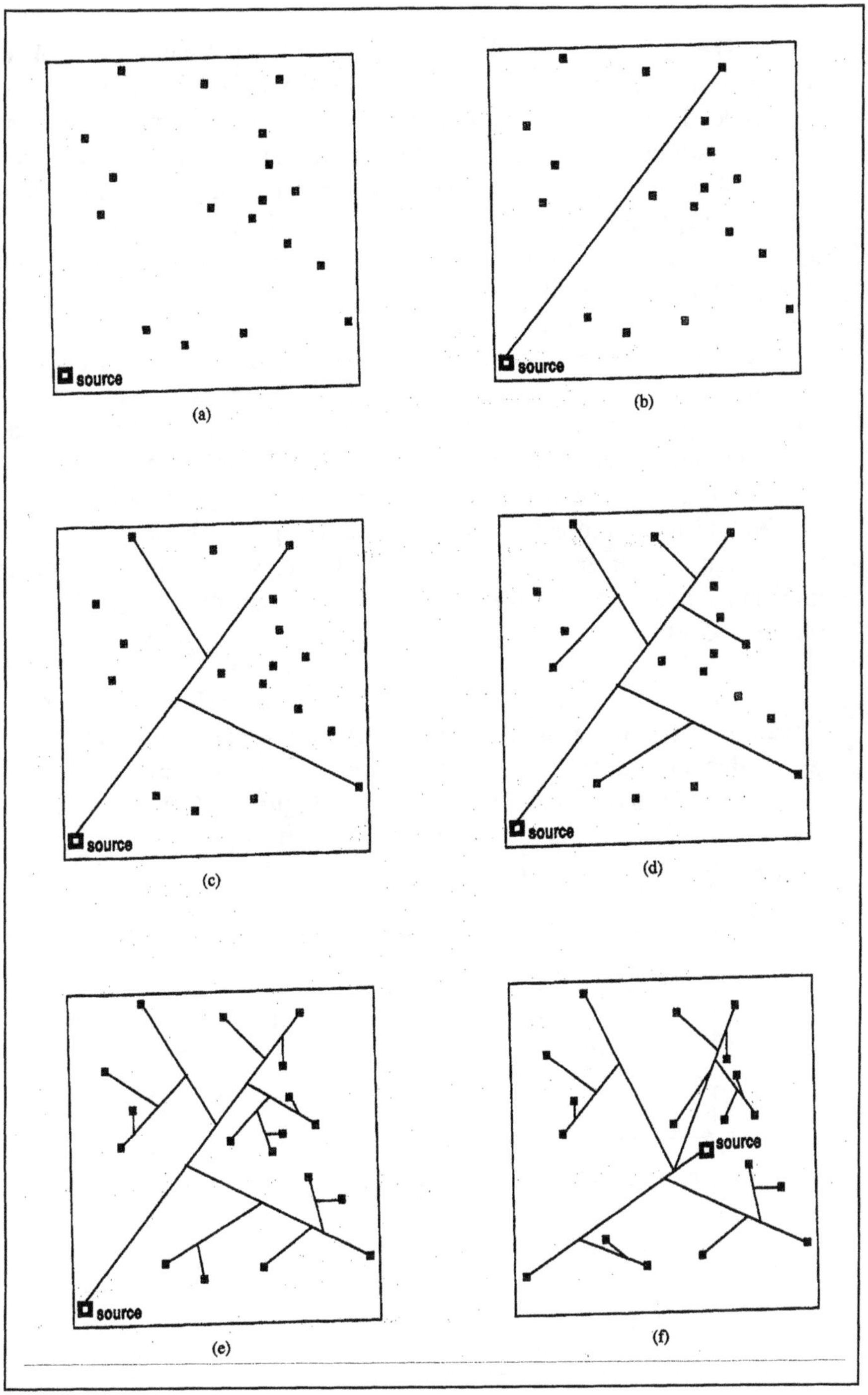

Figure 10-3. Constructing A Planar Equal Path Length Clock Tree [15].

10.2 Geometrical Embedding

The geometrical embedding transfers each branch of a planar equal path length clock tree to a set of rectilinear routing wires. The tree is embedded level by level in a bottom up order, starting from leaf branches, which are directly connected to the sinks. Two branches connected to the same parent node are called sibling branches. Two sibling branches are embedded together at each level. For each branch of the clock tree, we define the bounding box of the branch as the smallest rectangle that encloses the end points, as shown in Figure 10-4. When the bounding boxes of two sibling branches intersect, the intersection defines the base line of the two branches, as shown in Figure 10-4. A base line does not exist if the intersection of bounding boxes is degenerated to a point at the parent node, as shown in Figure 10-4(b). Evidently, if the base line exists and the embedding of sibling branches can share the base line, the total wire length will be decreased. A modified line-search algorithm has been developed to guarantee the planarity of the original planar equal path length clock tree [15]. Figure 10-5(b) shows the geometrical embedding result for the planar equal path length clock tree shown in Figure 10-5(a).

The clock trees, for a large-size example, are shown in Figure 10-6, with the original planar equal path length clock tree (left figure) and the geometrical embedding result (right figure). The clock source location can be flexibly selected in the tree construction. Figure 10-6 shows the results when the clock source is located at the center of the sinks; and Figure 10-7 shows the result of the clock source located at the boundary of the chip.

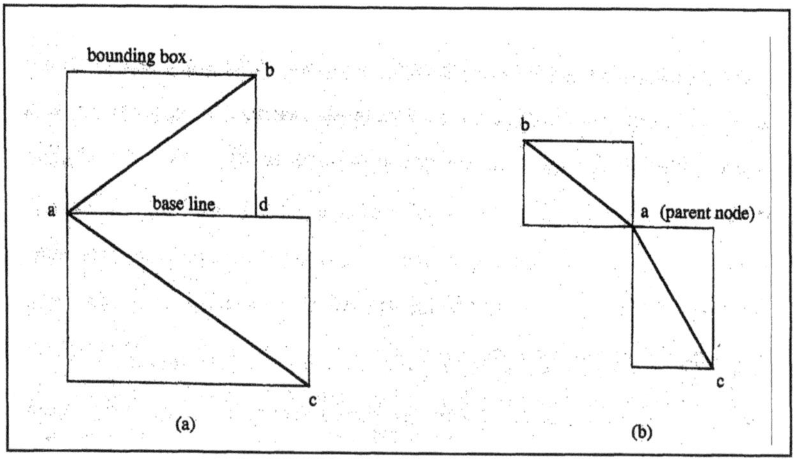

Figure 10-4. Base Line Sharing in Geometrical Embedding [15].

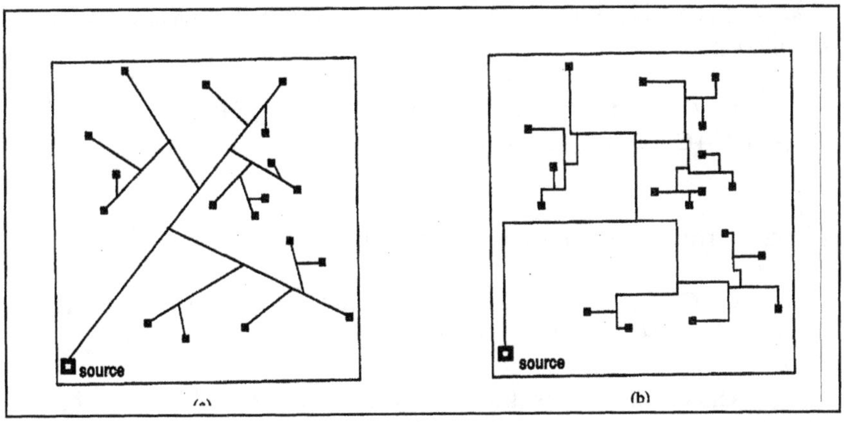

Figure 10-5. Geometrical Embedding for Planar Equal Path Length Clock Tree [15].

10.3 Skew-Bounded Refinement

One problem of the planar equal path length clock tree is that, even after the geometrical embedding, a large number of the total wire length will be produced, compared to the minimum Steiner tree topology. The reason is that we want to guarantee the equal path lengths from the clock source to all sinks. In most designs, the clock skew is not needed to be exactly zero, as long as it is within a tolerable skew bound. That means that if the equal path length clock tree has a very small clock skew, based on the delay calculation, it may still have the room to reduce the total wire length within the required clock skew bound. In order to make this improvement for the clock tree, a delay estimation model is necessary, so that we can continuously calculating the clock skew from the source to all sinks during the clock tree topology refinement.

A well-known model for the delay calculation is shown in Figure 10-8, where the clock tree is modeled with the resistance and capacitance for the sinks and interconnects. Based on this RC tree model, the delay from the source s to a sink t can be calculated in the following method. Multiply the resistance by the summation of the capacitance loads in the sub-tree rooted by this resistance, which is called an RC term. Then, from the source s to any sink, say, t_4 in Figure 10-8, the delay is estimated by the summation of all RC terms for the resistance branches along the path in the tree from *s* to t_4. The above method is also called the *Elmore Delay Modeling* [16,123]. For example, the delay from s to sink C_{t4} is calculated as follows:

$$R_d(C_d+C_1+C_2+C_3+C_4+C_5+C_6+C_7+C_8+C_{t1}+C_{t2}+C_{t3}+C_{t4})+R_1(C_1+C_2+C_3+C_4+C_5+C_6+C_7+C_8+C_{t1}+C_{t2}+C_{t3}+C_{t4})+R_3(C_3+C_4+C_5+C_6+C_{t3}+C_{t4})+R_5(C_5+C_6+C_{t4})+R_6(C_6+C_{t4}).$$

The delay from s to sink C_{t1} is also calculated as follows:

$$R_d(C_d+C_1+C_2+C_3+C_4+C_5+C_6+C_7+C_8+C_{t1}+C_{t2}+C_{t3}+C_{t4})+R_1(C_1+C_2+C_3+C_4+C_5+C_6+C_7+C_8+C_{t1}+C_{t2}+C_{t3}+C_{t4})+R_2(C_2+C_7+C_8+C_{t1}+C_{t2})+R_7(C_7+C_{t1}).$$

The clock skew is therefore defined between C_{t1} and C_{t4} based on the difference of above two path delays. We can get the worst-case clock skew between the path delays from the source to all sinks in the clock tree.

The refinement is iteratively applied to the clock tree by a cut-and-link operation. A long edge in the clock tree is removed, and the clock tree becomes two disjoint sub-trees. A skew constrained shorted path between two disjoint sub-trees is defined as the path with the smallest cost, subject to the constraint that the new tree connected by the path with two disjoint sub-trees still satisfies the clock skew bound. The algorithm to find a skew-bounded shortest path is a modification to the first k-shortest paths based on extending the Dijistra's algorithm for the shortest path [15]. The found skew-bounded shortest path is then added back to reconnect the two sub-trees. The skew-bounded shortest path is less than the path to be deleted; so the total wire length is reduced. In the worst case, the skew-bounded shortest path is the original path to be deleted. Once all the paths are exhaustively deleted and all the skew-bounded shortest paths are the same as the original paths to be deleted, the algorithm terminates with no hope to further reduce the tree's total wire length. Figure 10-9 shows the basic operation for the consecutively cutting and linking the clock tree in order to reduce the total wire length. Figures 10-10(a) and (b) show the starting and final tree topologies, respectively. The total wire length is significantly reduced in the final tree topology. Both trees satisfy the given skew bound of 0.25ns in this example.

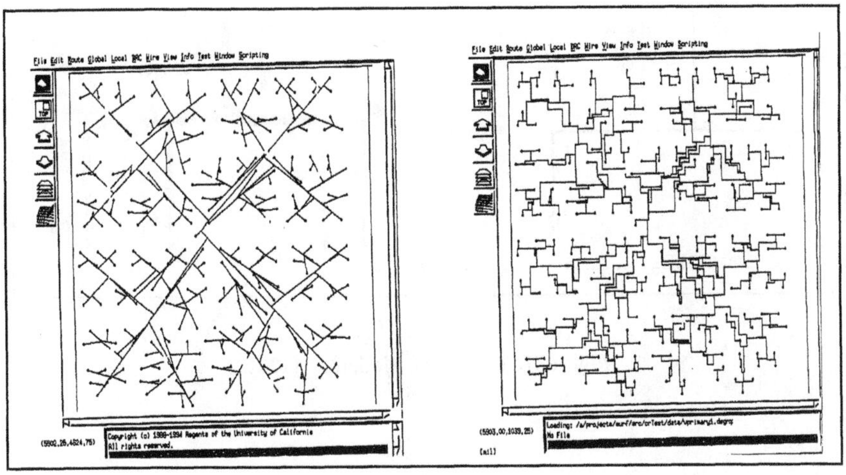

Figure 10-6. Clock Source Located at Center [15].

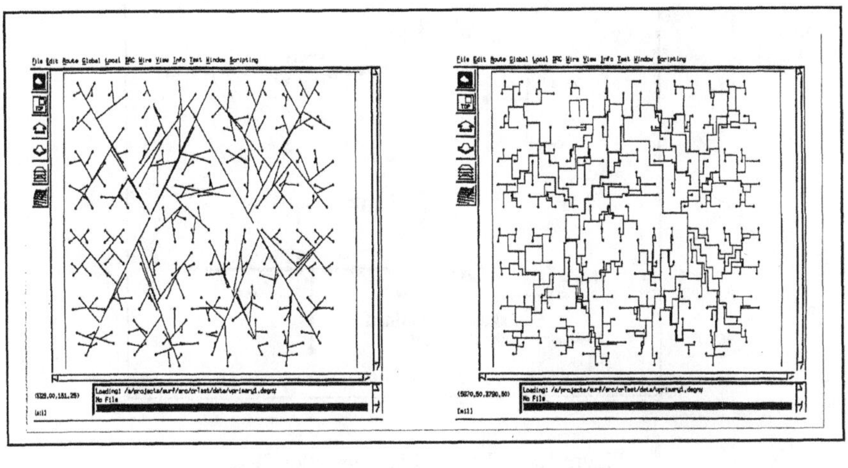

Figure 10-7. Clock Source Located at Boundary [15].

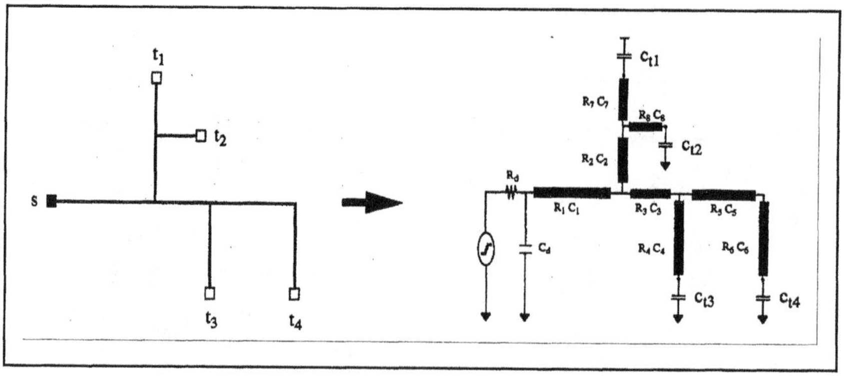

Figure 10-8. RC Tree Modelling [122].

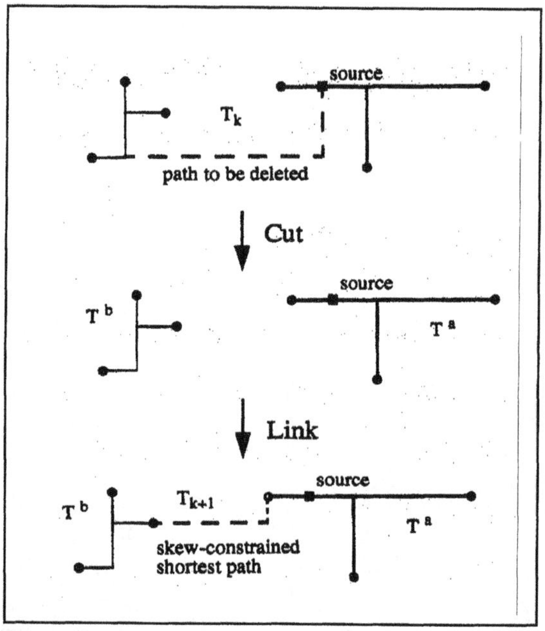

Figure 10-9. Cut-and-Link Operation [15].

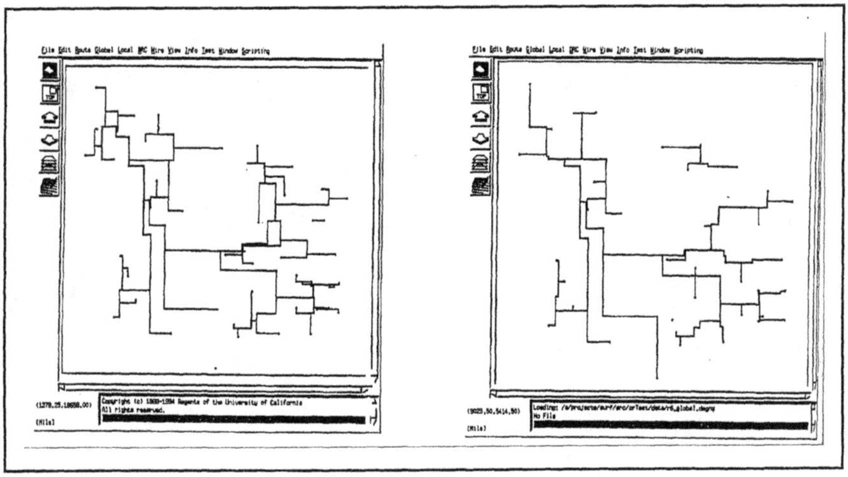

(a) (b)

Figure 10-10. Skew-bounded Clock Tree [15].

10.4 Wire Sizing of Clock Network

Wire sizing technique involves assigning different metal widths for the branches in the clock tree [29]. The advantage of the wire sizing is to reduce the clock delay instead of making the wires longer to match the delays. Figure 10-11 shows one example. C_2 has a larger capacitance load than C_1. So, even if we achieve the same path lengths from source s to two targets t_1 and t_2, we can still get different RC delays from s to t_1 and t_2. By making the path length from s to t_1 longer as shown in Figure 10-11(a), we can slow down the delay in the path s to t_1 to match the original longer delay in the path s to t_2. Another approach is to make the path s to t_1 with wider metal width, which reduces the metal resistance and therefore RC delay, as shown in Figure 10-11(b). Although both approaches achieve zero skew, the metal sizing techniques reduces the path delay, while the wire length increase approach increases the path delay. The third approach is to add more dummy capacitance loads on the C_1 side, which has a smaller load than C_2's, in order to balance the delays from s to t_1 and t_2.

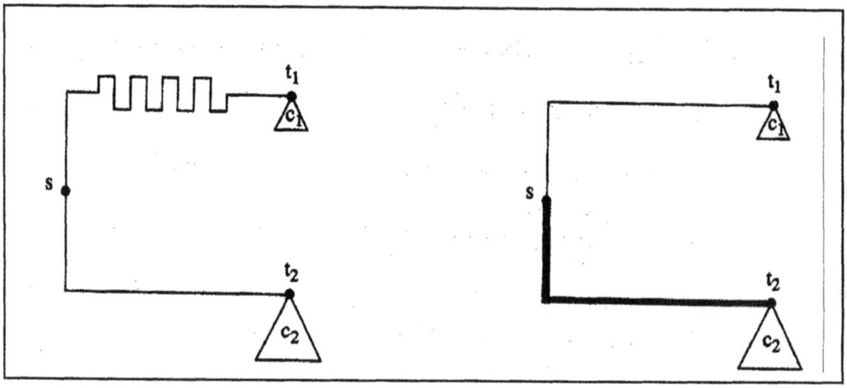

(a) (b)

Figure 10-11. Zero Skew Techniques [29].

An initial or rough wire-sizing algorithm for the clock tree has been proposed in [29]. It assumes all the branches in the clock tree have the minimum width. The algorithm starts to assign one branch at a time with the best feasible width that achieves the smallest clock skew. For all the clocked sinks in the tree, we select the one with the longest delay and always try to speed it up by sizing the branch as the ancestor branch of the slowest sink. Always sizing the branches in a top down order means that the branches are sized in the order from the clock source to the clocked sinks. The algorithm explores the clock tree from the source terminals using the breadth-first search. A priority queue is used to store all branches in the search wave frontier that are discovered but have not been explored. The branch in the queue, which is the ancestor branch of the slowest terminal with the largest path delay, is explored next. This branch is explored by assigning it the best feasible width for the largest skew reduction. The search wave frontier is then expanded to the children branches of this branch, by inserting children branches into the queue. Note that the slowest terminal is changed during this course. This process is repeated, until the parent branch of the slowest terminal has been explored, since further sizing of remaining branches would not help the reduction of the clock skew.

Figure 10-12 shows an example using the above algorithm to reduce the clock skew. The branches marked with "X" in the figure will be selected for the sizing in the next step. Again, the entire sizing process starts from the branches connected to the clock source and terminates when the slowest terminal cannot be sped up anymore after the sizing of the leaf branch that is connected to the slowest terminal in the tree.

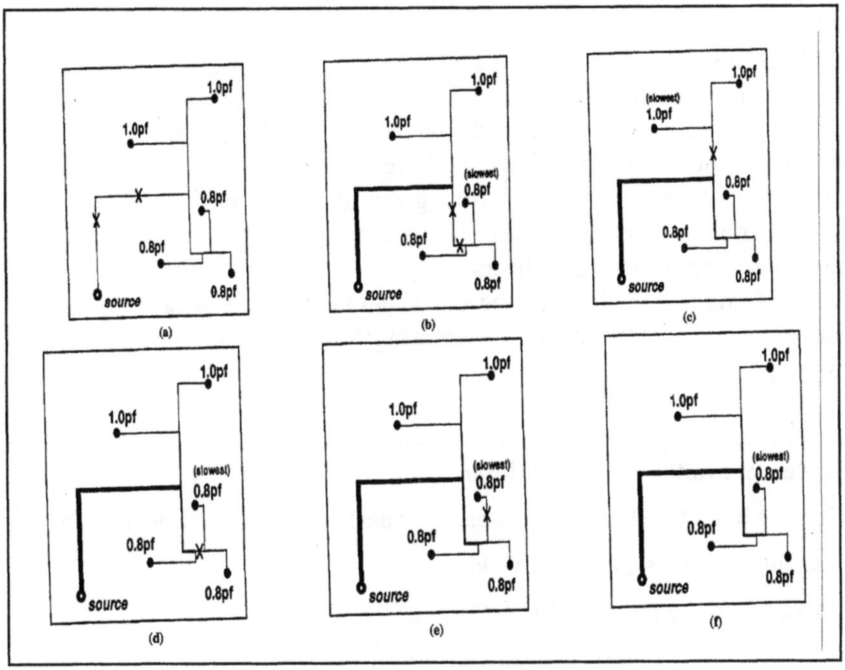

Figure 10-12. Clock Tree Sizing Example [29].

Figure 10-13 summarizes the above algorithm for initial wire sizing in the clock tree. Behind the algorithm, a delay estimation model is employed to estimate the path delays or the clock skew in each sizing step. The model can be as simple as the Elmore delay model in a RC tree topology, or can be a more accurate circuit simulator.

In general, the clock wire sizing is formulated as a least-squares estimation problem, with the root-mean-square error between paths delays linearly bounded by the clock skew [29]. Most algorithms for the least-squares estimation of non-linear parameters have centred on either of two methods [124]. In one method, the objective model is expanded as a Taylor series and corrections to parameters calculated in each iteration based on the assumption of local linearity. The other method is based on the modifications of the steepest-descent method. Both of these two methods have their disadvantages depending on applications. While the methods based on Taylor series suffer from a possible divergence of successive iterations, the steepest-descent based methods may have a very slow convergence to the optimum solution after the first little rapid iteration. The Gauss-Marquardt method tries to perform an optimal interpolation between the Taylor series method and the gradient-based method [124]. The optimum search is based upon the maximum neighbourhood in which the truncated Taylor series gives an adequate representation of the non-linear objective model. The Gauss-Marquardt method combines the best features of the previous two methods and may possibly avoid their limitations. It shares with the gradient methods the ability to converge from an initial guess quickly, and shares with the Taylor series methods the ability to converge rapidly after the vicinity of the converged values has been reached [124]. An application of the Gauss-Marquardt method on the clock network wire sizing optimization can be found in [29]. An uphill exploring for the optimal solution is also discussed [29].

Clock tree initial sizing algorithm

Input: a clock tree *T*, a source, a set of terminals and an equivalent circuit mod

Output: *T* with assigned branch widths.

```
Procedure ClockTreeInitialSizing(T) {
        Q = set of branches connected to the source;
        t_s = the slowest terminal of the clock tree;
        b = ancestor branch of t_s extracted from Q;
        s = skew of the clock tree;
        While (b Not Empty) {
                Assign b the feasible width, which results in the smallest s;
                For (each children branch b_i of b)
                        Insert b_i into Q;
                Update t_s;
                b = ancestor branch of t_s extracted from Q;
        }
}
```

Figure 10-13. Clock Tree Sizing Optimization Algorithm [29].

10.5 Summary

With the increasing complexity of clock tree construction, researches have been done in multiple areas to automate the clock routing task by CAD tools. This chapter collects our study in the planar clock routing and wire sizing algorithms. Many other references can be found on these topics [7,8,9,10,11,12,13,16,28,29]. Inserting the clock buffers in the clock trees for design automation to reduce the clock skew and power has also been discussed in papers [26,27,30,31]. Based on the author's knowledge, the research areas for CAD algorithms in clock distribution are as follows: (1) constructing the balanced clock trees by the routing, wire sizing, and buffering; (2) minimizing the power consumption for the clock distribution by routing and buffer insertion; and (3) doing the clock network synthesis by re-timing and non-zero local skew budgets [129,130]. Research works presented in recent years, which have not been included in this book, can be found in IEEE conferences on Integrated Circuits Design Automation and ICCAD.

Chapter 11 Clock Tree Design Flow in ASIC

This chapter shows the clock tree design flow in the timing optimised layouts based on the place and route CAD tools [61]. It shows one example of automation CAD tools used for the clock network design. The chapter is organized in six sections. Section 11.1 introduces the flow overview of the clock tree synthesis. Section 11.2 shows the detailed commands for clock buffer synthesis. Section 11.3 presents the commands to report the skew and clock tree topology. Section 11.4 describes the commands to route the clock net. Section 11.5 shows the commands to verify the clock skew after the clock net is routed.

11.1 Flow Overview

The overall design flow of the timing-driven place and route program for the application-specific integrated circuit (ASIC) chips using the standard cells is shown in Figure 11-1 [61]. The clock tree synthesis steps are shown in Figure 11-2 [61]. Non-critical high fan-out nets like scanning resets and enables need to be buffered. These have been placed in the flow so that they do not disturb critical clock nets as much. If performed after critical nets, the subsequent ECO placement run to buffer the non-critical nets may move the buffers on the critical nets. As shown in Figure 11-2, the gated CTS (GCTS) command synthesizes all levels of a buffered clock net. GCTS automatically processes the entire clock net, from the root net down to the clock pins or any synchronous points. Given the root net, GCTS traverses through all the gated sub-nets, determine the structure of buffer trees, and perform clock tree synthesis to minimize the overall skew. GCTS uses a more accurate delay/skew calculator, which enables the effective post clustering optimization, and built-in post-CTS optimization. GCTS also automatically generates the scripts for clock net routing.

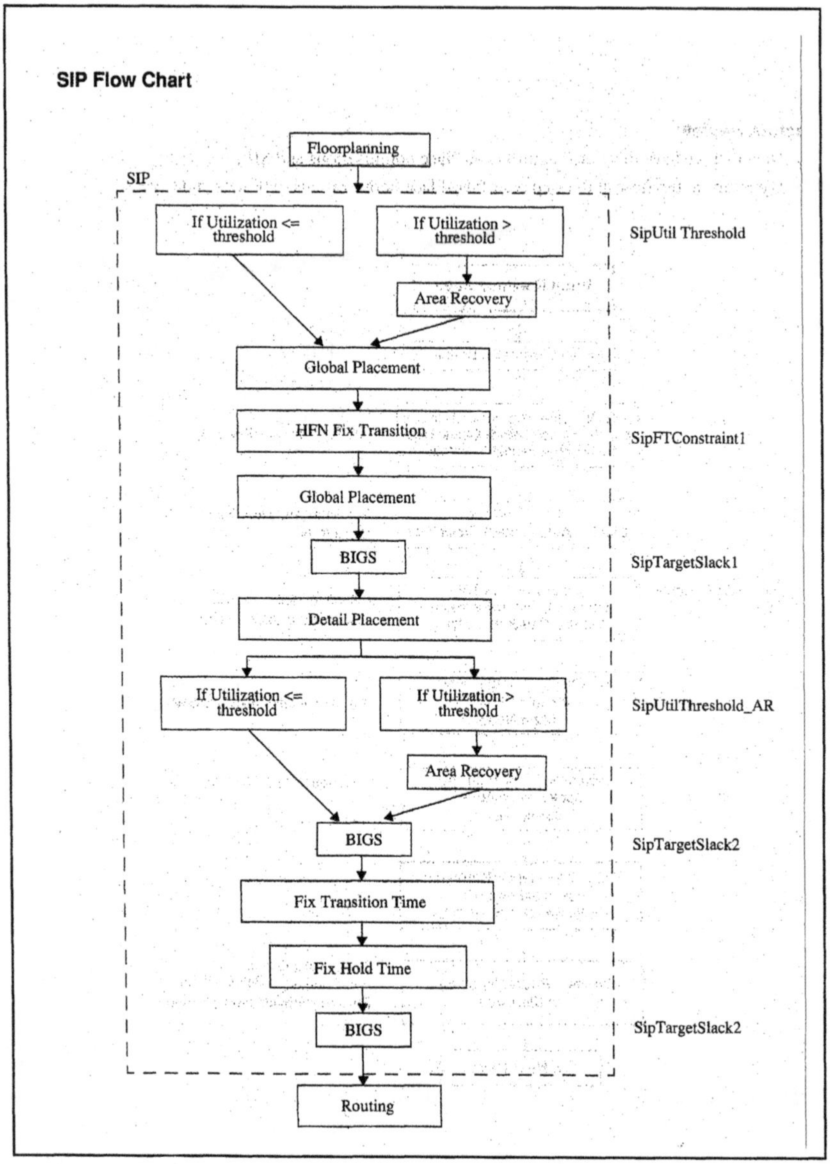

Figure 11-1: Timing Optimised Place and Route Flow [61].

The cell and buffer sizing step sizes the gates and buffers in the clock tree using the cell library in order to meet the required clock skew target. Gated or non-gated, buffered or un-buffered clock trees may be used with this clock buffer size optimization. This step is iterated with the ECO placement. This clock buffer sizing optimization should also be repeated after the clock is routed to see if there can be any further improvement on the clock skews. The level adjust step tries to adjust the connections of buffers inside a synthesized clock to further minimize the clock skew. This is an optional step, but it is recommended for improving a GCTS that drives macro clocks or gated clock trees.

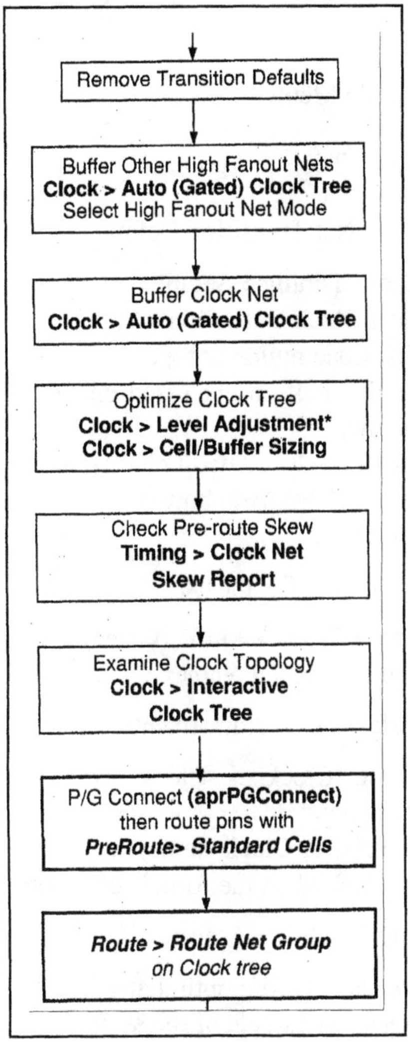

Figure 11-2: Clock Tree Synthesis Flow [61].

11.2 Gated Clock Tree Synthesis

In this step, all levels of a gated clock net are synthesized to minimize the overall clock skew. The following are detailed procedures for gated clock tree synthesis in a timing-driven Place and Route tool [61].

1. Load the Apollo Menu:
   - Choose Tools > Apollo II;
2. Remove Clock Net Transition Default:
   - Choose Timing > Setup;
   - Select Clock Definition tab;
   - Click Load button at top;
   - Double click on clk listed on the List of Clock Sources window;
   - Deselect Select default clock transition time to button;
   - Select Modify/Add button;
   - Hide form.
3. Perform Clock Tree Synthesis:
   - Choose Clock > Auto. (Gated) Clock Tree …;
   - Type clk for Net Names.
4. Perform Buffers Level Adjustment:
   - Choose Clock > Clock Tree Optimization – Level Adjustment;
   - Type clk for Clock Name;
   - Click OK when the form is complete.
5. Examine Log Window:
   - Look for skew before optimizing clock and afterwards. Figure A-3 shows an example of the skew log file.
6. Perform Cell/Buffer Sizing:
   - Choose Clock > Clock Tree Optimization –Cell/Buffer Sizing;

- Type clk for Clock Name;
- Click OK when the form is complete;
- Examine Log File for skew numbers.

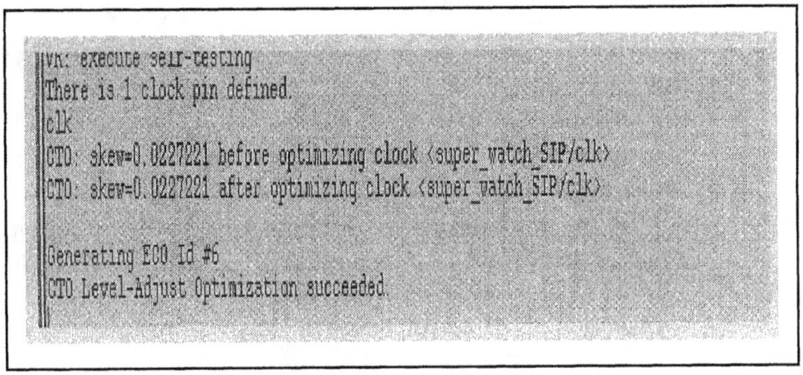

```
VN: execute self-testing
There is 1 clock pin defined.
clk
CTO: skew=0.0227221 before optimizing clock <super_watch_SIP/clk>
CTO: skew=0.0227221 after optimizing clock <super_watch_SIP/clk>

Generating ECO Id #6
CTO Level-Adjust Optimization succeeded.
```

Figure 11-3: Log file [61].

11.3 Clock Skew and Topology Reports

Before routing the design, we need to verify the clock skew and timing results based on the following procedures [61].

7. Load the Apollo Menu:
   - Choose Timing > Clock Net Skew Report;
   - Type clk for Clock Name;
   - Click OK when the form is complete.
8. Save Pre-route Clock Skew Report:
   - Click save as in the text window;
   - Type the skew report file name.
9. Generate Timing Report:
   - Choose Timing > Verify;
   - Click OK.
10. Save Pre-route Timing Report:
    - Click Save As in the pop-up text window;

- Type the timing report file name;
- Click OK.

11. Save Current Cell:

    - Cell > Save.

12. Examine the Clock Topology:

    - Launch the clock tree browser: Clock > Interactive Clock Tree;
    - Examine the layout of the clock tree using the options in the interactive CTS form. The Browser Panel displays various levels of the clock tree. The Info Panel displays the skew and path information. The Cross Probe Panel shows the physical layout of the logical clock tree. The ECO Panel allows adding cells or re-parenting the selected cells.

11.4 Route the Clock Net

It is recommended that the clock nets be routed first so that there is enough space for more direct routing. One criterion for clock routing is to minimize the clock skew. The Balanced Option must be selected in the Route Common Options Form as follows [61].

13. Pre-route Standard Cells:

    - Choose Pre-route > Standard Cells;
    - Use the default options on the form;
    - Click OK.

14. Using Route Net Group:

    - Set Route Options to minimize the skew. Choose Route > Route Common Options. Select Balanced for Clock Routing. Click OK when the form is complete;
    - Route Clock Net. Choose Route > Route Net Group. Type clk.* for Net Names;
    - Click OK when the form is complete.

15. Save the Current Cell:

    - Choose Cell > Save.

16. Route the Design:

- Set Route Options to meet timing constraints. Select Timing Driven and click OK when the form is complete;
- Route the design. Choose Route > Auto Route.

17. Save and Close the Current Cell:

- Choose Cell > Close;
- Click OK.

11.5 Verify the Clock Skew

After routing is done, check the clock skew and timing and compare to the values obtained before the routing.

18. Generate the Clock Net Skew Report:

- Choose Timing > Clock Net Skew Report;
- Type clk for Clock Name;
- Click OK when the form is complete;
- Save the Post-route Clock Skew Report.

19. Propagate the Clocks Using Actual Clock Skew:

- Choose Timing > Setup;
- Select Clock Definition Tab;
- Select Use Actual Clock Skew button;
- Click on Hide button.

20. Generate the Timing Report:

- Choose Timing > Verify;
- Click OK.

21. Save the Post-route Timing Report:

- Click save as in the text window;
- Type the timing report file name;
- Click OK

11.6 Summary

The place and route CAD tools are widely used in Application Specific Integrated Circuits (ASIC) design, which are based on the standard cells for basic logic circuits. Since the time to the market is critical for ASIC products, the clock buffers placement and clock trees routing become the automated features in the modern place and route tools as described in this chapter. The designers need to specify the insertion delay or skew budget from the clock source to clocked elements in the standard cells layout, and the tool will based on the required delay or skew to decide the clock buffer levels and routing structures for standard cells layout. Notice that the insertion delay or skew budget is decided based on the detailed full-chip clock skew and clock delay analysis. In addition, the clock tree constructed by the place and route tool needs to be verified using the transistor-level circuit simulation, since most of these tools will estimate the delays in simple delay equations, which may miss the accuracy such as the rise time of input clocks to clocked elements.

Glossary

ASIC: Application specific integrated circuits. An ASIC chip is usually designed by the CAD flow to speed up the design cycle.

Balanced Clock Tree: The delays from the root of the clock tree to the leaves are almost the same. An example of a balanced clock tree is the recursive H-tree structure.

Clock Distribution: The task of distributing the clock signal across the chip. The main objective is to minimize the clock skew, which is the difference among the clock arrival times at various clocked elements.

Clock Buffer: A set of basic inverter circuits inserted in the clock tree for keeping sharp rise and fall times of the clock signal. Generally, the clock buffers increase the clock skew because of the variations in process, supply voltage, and temperature across the chip.

Clock Buffer Design: The task of determining the levels and device sizes of clock buffers for the clock distribution.

Clock Buffer Tuning: The device sizing in clock buffers to reduce the clock skew across the chip.

Clocked Elements: The set of logic elements in the chip that need clock signals for synchronization. The clocked elements discussed in this book includes latches, flip-flops, and domino circuits.

Clock Gating: The technique to stop the clock from clock buffers by a gating signal. It reduces wasted power whenever the clock signal is not required for the sub-circuits.

Clock Grid: The metal lines shorted in the grid structure for clock distribution. It is one of the common clock topologies used in high-speed design.

Clock Inaccuracy: The shifting of clock edges from their ideal positions. Clock inaccuracy is attributable to clock skew and clock jitter.

Clock Jitter: The clock edges shifting from their ideal positions in different clock cycles. The supply voltage variation in clock buffers is one of main sources of clock jitter.

Clock Nets: The nets on the chip for the clock signals. A chip may have multiple clock nets with different frequencies and phases.

Clock Routing: The task of connecting clock buffers and clocked elements.

Clock Skew: The delay differences across the clock network on the chip from the clock source to multiple clocked sinks. Clock skew is a significant component contributing to the required minimal clock cycle time in order to avoid the synchronization problems.

Clock Trunk: One or several wide clock trunks across the chip. The neighbouring clocked elements are connected to these wide trunks.

Clock Phase: Defined by zero and one-voltage levels based on clock signal waveforms in different time domains or edge positions.

Clock Simulation: The task of analysing the clock network for the clock skew and clock path delays.

Clock Skew Analysis: The same as the clock simulation.

Clock Skew Budget: The upper limits of clock skews allowed in each localized clocking region or a global skew budget across the chip.

Clock Synthesis: The task of determining the clock buffer levels and sizes.

Critical Nets: The most critical nets to determine the chip performance. Examples include the clock network and power network.

D1/D2 Dominos: Two types of the most commonly used domino circuits. D1 domino has a clocked node at the NMOS evaluation stack. D2 domino does not have the clocked node. D2 is usually faster than D1 but has a potential risk in timing.

Distributed RC: The interconnect resistance is also included in the circuit modelling in addition to the interconnect capacitance for the detailed circuit simulation.

DLL: An acronym for delay-locked loop. It contains a chain of inverters and control logic for delay adjustment based on the phase match to a reference clock at the inputs.

De-Skewing Buffer: The same function as DLL.

Differential Clocks: Two inverse-phase switched clock signals, usually in low-voltage swings. Differential clocks has the advantages of low power and common-mode noises cancellation.

Dynamic Timing Analysis: The circuit simulation in detailed inputs of signal waveforms. SPICE-like transistor level circuit simulation is one kind of the dynamic timing analysis.

Domino Circuits: The logic elements, not data storage elements, implemented in two clock phases: pre-charge phase and evaluation phase. The domino circuit provides high speed because it uses fewer PMOS devices. Also, the pipelines are possible in the same clock cycle for the ripple effect in a long domino implemented logic stage.

Gated Clock: The clock signal is gated by a stop signal at the clock buffer (AND gate) to reduce the wasted clocking power.

RC Extraction: The process to build the RC models of interconnects in VLSI layout.

Fall Time: The time interval it takes a waveform (signal) to fall from 90%Vdd to 10%Vdd voltage levels.

Fan-out: The number of receivers that a driver is connected in a netlist.

Flip Flop: The data storage element trigged by the clock rising or falling edges. The circuit consists of two latches in a master and slave sequence.

Gate Netlist: The description of logic gates and interconnections to implement the logic function. The gate netlist is written in Verilog code or SPICE-like netlist.

Global Clock Driver: The highest-level clock buffer in an on-chip clock distribution network. It comes after the phase-locked loop (clock generator) and drives the global clock network.

Global Clock Network: The on-chip clock distribution is usually implemented with a global clock network followed by many small-scale local clock networks.

H Clock Tree: The recursive H structures for the clock distribution. It is the best-known structure for minimizing the clock skew, outperforming other clock routing structures such as the clock grid or clock trunk.

High Fan-out Nets: Nets with high fan-outs. Examples of high fan-out nets are clock nets, reset nets, and address nets.

Hold Time: The hold time for a sequential cell is the minimum length of time the data-input signal must remain stable after the active edge of the clock to ensure correct functioning of the cell. The hold constraint value is measured as the interval between the clock signal crossing 50% of Vdd and the data signal crossing 50% of Vdd.

Latch: The data storage element triggered by the high or low voltage levels of the clock signal.

Local Clock Network: The small-scale clock network extended from the on-chip global clock network. The clocked elements are connected within local clock networks.

Max Speed Path: The setup time violations in the circuit due to long logic delays across the circuit. The max speed paths, which cause the setup time violations, have to be fixed by reducing the logic path delays or increasing the clock path delays.

Min Speed Path: The hold time violations in the circuit due to the short logic delays across the circuit. The min speed paths, which cause the hold time violations, have to be fixed by increasing the logic path delays or reducing the clock path delays.

Metastability: The situation of uncertain data logic values for a latch or flip-flop data storage element. Metastability is caused by violations of the minimum setup or hold time requirements for the latch or the flip-flop.

On-die Clock Shrink: A circuit technique to detect the timing violations in the silicon test by adjusting the clock edges.

PLL: An acronym for phase-locked loop. The phase-locked loop is used to make the phase of the output clock align with the phase of the reference input clock. The phase-locked loop is implemented using the VCO (voltage controlled oscillator), adjusting the phases, and multiplying the reference clock frequency in the output clock.

Propagation Delay: The time interval between the input stimulus crossing 50%Vdd and the output crossing 50%Vdd.

Pulse Width: The length of time between the leading and trailing edges of a pulse waveform. Minimum pulse width high (minpwh) is measured as the interval between the rising edge of the signal crossing 50%Vdd and the falling edge of the signal crossing 50%Vdd. Minimum pulse width low (minpwl) is measured as the interval between the falling edge of the signal crossing 50%Vdd and the rising edge of the signal crossing 50%Vdd.

PVT Variations: The variations across the chip of the transistor process (gate length, gate width, etc.), the supply voltage, and the junction temperature. The PVT variations have to be carefully modelled in order to simulate the circuit functions correctly.

Rise Time: The time interval it takes the waveform (signal) to rise from 10%Vdd to 90%Vdd voltage levels.

Recovery Time: The minimum length of time that the active-low set or reset signal must remain high before the active edge of the clock in order to ensure correct functioning of the cell. Recovery constraint values are measured as the interval between the set or reset signal crossing 50%Vdd and the clock signal crossing 50%Vdd.

Reference Clock: The input clock to the PLL or de-skewing buffer. The output clock edges from the PLL or de-skewing buffer will align with this input reference clock.

Standard Cells: The design style of using pre-designed cells for basic logic functions and storage elements.

Setup Time: The minimum length of time the data-input signal must remain stable before the active edge of the clock (or other specified signal) to ensure the correct functioning of the sequential cell. Setup-constraint values are measured as the interval between the data signal crossing 50%Vdd and the clock signal crossing 50%Vdd.

Slack Graph: A compact representation of the timing constraints. The slack graph is a directed graph where each pin in the netlist is represented by a node, and each signal path between pins by an edge.

Static CMOS Circuit: The traditional logic elements, not data storage elements, using a pair of NMOS and PMOS devices.

Static Timing Analysis: The violation checks for setup and hold time constraints in the circuit. It is performed based on the slack graph and timing models of logic elements and macro blocks.

Timing Constraints: The timing constraints of sequential circuits (flip-flops and latches) defining the minimum time intervals between the data signals and clocks. They specify signals must be ready in order to ensure the correct functioning of sequential circuits. Timing constraints includes the setup time, the hold time, the recovery time, and the minimum pulse width.

Time Borrowing: The technique of one stage in the logic borrowing time from the next stage logic across latches.

Transition Time: The time needed for the input signal to go from 10%Vdd to 90%Vdd voltage levels. The output transition time is defined as the time needed for the output signal to go from 10%Vdd to 90%Vdd voltage levels.

Timing Model: The modelling language to describe the timing behaviour of functional blocks.

Voltage Controlled Oscillator: A key component in the phase-locked loop to generate the clock waveform and clock frequency.

Reference

1. J. P. Fishburn, "Clock Skew Optimization," IEEE Transactions on Computers, Vol. 39, No. 7, pp. 945-951, 1990.

2. H. Bakoglu, Circuits, Interconnections, and Packaging for VLSI, Addison-Wesley Publishing Company, 1987.

3. F. Wann and M. A. Franklin, "Asynchronous and Clocked Control Structures of VLSI-Based Interconnection Networks," IEEE Transactions on Computers, Vol. 32, No. 3, pp. 284-293, 1997.

4. G. Gerosa, S. Gary, C. Dietz, D. Pham, K. Hoover, J. Alvarez, H. Sanchez, P. Ippolito, T. Ngo, S. Litch, J. Eno, J. Golab, N. Vanderschaaf, and J. Kahle, "A 2.2W, 80MHz Superscalar RISC Microprocessor," IEEE Journal of Solid-State Circuits, Vol. 29, No. 12, pp. 1440-1454, 1994.

5. J. Montanaro, R. T. Witeck, K. Anne, A. J. Black, E. M. Cooper, D. W. Dobberpuhl, P. M. Donahue, J. Eno, A. Farell, G. W. Hoeppner, D. Kruckemyer, T. H. Lee, P. Lin, L. Madden, D. Murray, M. Pearce, S. Santhanam, K. J. Synder, R. Stephany, and S. C. Thierauf, "A 160MHz 0.5W CMOS RISC Microprocessor," Proc. of IEEE International Solid-State Circuits Conference, pp. 214-215, 1996.

6. D. W. Dobberpuhl and R. T. Witek, "A 200 MHz 64b Dual-Issue CMOS Microprocessor," Proc. of IEEE International Solid-State Circuits Conference, pp. 106-107, 1992.

7. B. Jackson, A. Srinivasan, and E. S. Kuh, "Clock Routing for High-Performance ICs," Proc. of 27th Design Automation Conference, pp. 573-579, 1990.

8. A. B. Kahng, J. Cong, and G. Robins, "High-Performance Clock Routing Based on Recursive Geometric Matching," Proc. of 28th Design Automation Conference, pp. 322-327, 1991.

9. R. S. Tsay, "An Exact Zero-Skew Clock Routing Algorithm," IEEE Transactions on Computer-Aided Design, Vol. 12, No. 3, pp. 242-249, 1993.

10. T. H. Chao, Y. C. Hsu, J. M. Ho, K. D. Boese, and A. B. Kahng, "Zero Skew Clock Net Routing," IEEE Transactions on Circuits and Systems, Vol. 39, No. 11, pp. 799-814, 1992.

11. D. Boese and A. B. Kahng, "Zero Skew Clock Routing Trees with Minimum Wirelength," Proc. of 5th IEEE International Conference on ASICs, pp. 17-21, 1992.

12. M. Edahiro, "A Clustering-Based Optimization Algorithm in Zero-Skew Routings," Proc. of 30th ACM/IEEE Design Automation Conference, pp. 612-616, 1993.

13. C. Chou and C. K. Cheng, "Wire Length and Delay Minimization in General Clock Net Routing," Digest of Technical Papers of IEEE International Conference on Computer Aided Design, pp. 552-555, 1993.

14. Q. Zhu and W. W.-M. Dai, "Perfect-Balance Planar Clock Routing with Minimal Path-Length," Digest of Technical Papers of IEEE International Conference on Computer Aided Design, pp. 473-476, 1992.

15. Q. Zhu and W. W.-M. Dai, "Planar Clock Routing for Chip and Package Co-Design," IEEE Transactions on VLSI Systems, Vol. 4, No. 2, pp. 210-226, June 1996.

16. A. B. Kahng and C.-W. A. Tsao, "Low-Cost Single-Layer Clock Trees with Exact Zero Elmore Delay Skew," Digest of Technical Papers of IEEE International Conference on Computer Aided Design, pp. 213-218, 1994.

17. Q. Zhu and W. W.-M. Dai, "Chip and Package Co-Design for Clock Networks," Proc. of IEEE Multichip Modules Conference, pp. 160-163, 1996.

18. Q. Zhu and S. Tam, "Package Clock Distribution Design Optimization for High-Speed and Low-Power VLSIs," IEEE Transactions on CPMT/advanced Packaging, Vol. 20, No. 1, pp. 56 – 63, Feb. 1997.

19. E. G. Friedman, edited, Clock distribution network in VLSI circuits and systems, IEEE Press, 1995.

20. A. Chandrakasan, W. J. Bowhill, and F. Fox, edited, Design of High-Performance Microprocessor Circuits, IEEE Press, 2000.

21. T. Gneiting and I. P. Jalowiecki, "Influence of Process Parameter Variations on the Signal Distribution Behavior of Wafer Scale Integration Devices," IEEE Trans. on Components, Packaging, and Manufacturing Technology – Part B, Vol. 18, No. 3, pp. 424-430, Aug. 1995.

22. D. C. Keezer and N. Nigam, "A Comparative Study of Clock Distribution Approaches for WSI," Proc. IEEE Int. Conf. on Wafer Scale Integration, pp. 243-251, 1993.

23. D. C. Keezer and V. K. Jain, "Design and Evaluation of Wafer Scale Clock Distribution," Proc. IEEE Int. Conf. on Wafer Scale Integration, pp. 168-175, 1992.

24. C. Michael and M. Ismail, Statistical Modeling for Computer-aided Design of MOS VLSI Circuits, Kluwer Academic Publishers, 1992.

25. M. J. B. Bolt, Rocchi, and Engel, "Realistic Statistical Worst Case Simulations of VLSI Circuits," IEEE Trans. on Semiconductor Manufacturing, Vol. 4, No. 3, pp. 193-198, Aug. 1991.

26. M. Shoji, "Elimination of Process-dependent Clock Skew in CMOS VLSI," IEEE Journal of Solid-State Circuits, Vol. SC-21, No. 5, pp. 875-880, 1986.

27. J. G. Xi and W.-M. Dai, "Buffer Insertion and Sizing under Process Variations for Low Power Clock Distribution," Proc. of 32nd Design Automation Conference, pp. 491 – 496, 1995.

28. S. Pullela, N. Menezes, J. Omar, and L.T. Pillage, "Reliable Non-zero Skew Clock Trees Using Wire Width Optimization," Proc. of 30th Design Automation Conference, pp. 165-170, 1993.

29. Q. Zhu and W. W.-M. Dai, "High Speed Clock Network Sizing Optimization based on Distributed RC and RLC Interconnect Models," IEEE Transactions on Computer-

Aided Design of Integrated Circuits and Systems, Vol. 15, No. 9, pp. 1106-1118, 1996.

30. N. Hedenstierna and K. O. Jeppson, "CMOS Circuit Speed and Buffer Optimization," IEEE Transactions on CAD, pp. 270-281, 1987.

31. B. A. Richman, J. E. Hansen, and K. Cameron, "A Deterministic Algorithm for Automatic CMOS Transistor Sizing," IEEE Journal of Solid-State Circuits, pp. 522 – 526, April 1988.

32. L. Benini, A. Bogliolo, and G. D. Micheli, "A Survey of Design Techniques for System-Level Dynamic Power Management," IEEE Trans. on Very Large Scale Integration (VLSI) Systems, Vol. 8, No. 3, pp. 299 – 316, 2000.

33. L. Geppert and T. S. Perry, "Transmeta's Magic Show," IEEE Spectrum, pp. 26 – 33, May 2000.

34. Q. Zhu and M. Zhang, "Low-Voltage Swing Clock Distribution Scheme," IEEE International Symposium on Circuits and Systems, pp. IV-418-421, May 2001.

35. B. Gunning, etc, "A CMOS Low-Voltage-Swing Transmission-Line Transceiver," IEEE International Solid-State Circuits Conference, pp. 58-61, 1992.

36. Rambus - Architectural Overview, Rambus Inc., Mountain View, CA, 1993.

37. Bellaouar, I. S. Abu-Khater, and M. I. Elmasry, "An Ultra-Low-Power CMOS On-Chip Interconnect Architecture," IEEE Symposium on Low Power Electronics, pp. 52-53, 1995.

38. Y. Nakagome, K. Itoh, M. Isoda, K. Takeuchi, and M. Aoki, "Sub-1-V Swing Internal Bus Architecture for Future Low-Power ULSI's," IEEE Journal of Solid-State Circuits, Vol. 28, No. 4, pp. 414-419, April 1993.

39. H. Zhang, V. George, and J.M. Rabaey, "Low-Swing On-Chip Signaling Techniques: Effectiveness and Robustness," IEEE Transactions of Very Large Scale Integration, Vol. 8, No. 3, pp. 264 – 272, 2000.

40. P. Ta and K. Do, "A Low Power Clock Distribution Scheme For Complex IC System," Fourth Annual IEEE International ASIC Conference and Exhibit, pp. 5.1-5.4, 1991.

41. H. Kojima, S. Tanaka, and K. Sasaki, "Half-Swing Clocking Scheme for 75% Power Saving in Clocking Circuitry," Symposium on VLSI Circuits Digest of Technical Papers, pp.23-24, 1994.

42. H. Kojima, S. Tanaka, and K. Sasaki, "Half-Swing Clocking Scheme for 75% Power Saving in Clocking Circuitry," IEEE Journal of Solid-State Circuits, Vol. 30, No. 4, pp. 432-435, April 1995.

43. E. D. Man and M. Schobinger, "Power Dissipation in the Clock System of Highly Pipelined ULSI CMOS Circuits," International Low Power Design Workshop, pp. 133-138, 1994.

44. T. G. Noll and E. D. Man, "Pushing the Performance Limits Due to Power Dissipation of Future ULSI Chips," Proceedings of ISCAS-92, pp. 1652-1655, 1992.

45. M. Bazes, "Two Novel Fully Complementary Self-Biased CMOS Differential Amplifier," IEEE Journal of Solid -State Circuits, Vol. 26, No. 2, pp. 165-168, Feb 1991.

46. P. E. Allen and D. R. Holberg, CMOS Analog Circuit Design, Oxford University Press Inc., Chapter 7, 1987.

47. E. G. Friedman and S. Powell, "Design and Analysis of a Hierarchical Clock Distribution System for Synchronous Standard Cell/Macrocell VLSI," IEEE Journal of Solid-State Circuits, Vol. SC-21, No. 2, , pp. 240-246, April 1986.

48. G.M. Blair, "Skew-Free Clock Distribution for Standard-Cell VLSI Designs," Proceedings of IEEE, Vol. 139, No. 2, pp. 265-267, April 1992.

49. Q. Zhu and T.W. Chan, "Delay/Slope Budgeting for Clock Buffer Cell Design," The 8th IEEE International Conference on Electronics, Circuits and Systems, ICECS-2001, pp. 417-420, Sept. 2001.

50. G. Anderson, Optim Manual, Intel Corporation, 1999.

51. Q. Zhu and T.W. Chan, "Local Clock Distribution Study in Pinecrest," Intel Corporation, 2000.

52. HSPICE Manual, Avant! Corporation, April 1999.

53. S. Duvall, "Within-Die Variation Design Guidelines," Intel Corporation, 1999.

54. N. Hakim and S.G. Duvall, "Statistical Circuit Simulation of Within-Die Variations," Intel Corporation, 2000.

55. B. Woodruff and I. Gur-Arie, "Neutrino Clock Tree Synthesis for Timna North," Intel Corporation, 2000.

56. Q. Zhu, "Clock Buffer Cell Delay/Slope Budgeting and Design," Intel Corporation, 2001.

57. Q. Zhu and A. Waizman, "Optimization and Analysis Techniques for High Performance Microprocessor Power Network Design," Intel Corporation, 2001.

58. J. Barkatullah and T. Fletch, "Willamette Clocking Methodology," Intel Corporation, 2000.

59. "TSMC 0.25μm Enhanced Process (C025E) SAGETM Standard Cell Library," Artisan Component Inc., Jan. 1999.

60. Q. Zhu, "PIO Hold Time Simulation and Fixing in Viper-D," Chameleon Systems Inc., July 2001.

61. "Apollo Timing Optimized Layout Training Guide," Avant! Inc, Jan. 2001.

62. "PrimeTime Modeling User Guide," Synopsys Inc., Oct. 1999.

63. P. Aycinena, "Focus Report – Timing and Power Analysis," Integrated System Design Magazine, pp. 56-63, June 2001.

64. N. A. Kurd, J. S. Barkatullah, R. O. Dizon, T. O. Fletcher, and P. D. Madland, "A Multi-Gigahertz Clocking Scheme for the Pentium 4 Microprocessor," IEEE Journal of Solid-State Circuits, Vol. 36, No. 11, pp. 1647-1653, Nov. 2001.

65. G. Pham, "Sequential Machine," Intel Corporation, 2000.

66. "Domino Circuit Design Tutorial," Intel Corporation, 1995.

67. Q. Zhu, M. Lingareddy, and T. W. Chan "A New Method of Domino Low-Vt Design for Noise Protection," Intel Corporation, 2001.

68. V. De, Y. Ye, S. Narendra, J. Kao, D. Somasekhar, R. Nair, and S. Borkar, "Techniques for Leakage Power Reduction," Design of High-Performance Microprocessor Circuits, ed. by A. Chandrakasan, W. J. Bowhill, and F. Fox, IEEE Press, 1999, Chap. 3.

69. J. Kao, "Dual Threshold Voltage Domino Logic," Proc. of 25th European Solid State Circuits Conference, pp. 409-414, Sep. 1999.

70. S. Mutoh, T. Douseki, Y. Matsuya, T. Aoki, S. Shigematsu, and J. Yamada, "1-V Power Supply High-Speed Digital Circuit Technology with Multithreshold-Voltage CMOS," IEEE Journal of Solid-State Circuits, Vol. 30, No. 8, , pp. 847-854, Aug. 1995.

71. J. Kao, A. Chandrakasan, and D. Antoniadis, "Transistor Sizing Issues and Tools for Multi-Threshold CMOS Technology," Proc. of 34th Design Automation Conference, pp. 409-414, June 1997.

72. L. Wei, Z. Chen, M. Johnson, and K. Roy, "Design and Optimization of Low Voltage High Performance Dual Threshold CMOS Circuits," Proc. of 35th Design Automation Conference, pp. 489-494, June 1998.

73. R. H. Krambeck, "High Speed Compact Circuits with CMOS," IEEE Journal of Solid-State Circuits, Vol. SC-17, No. 3, pp. 614-619, June 1982.

74. K. Bernstein, "Basic Logic Family," Design of High-Performance Microprocessor Circuits, ed. by A. Chandrakasan, W. J. Bowhill and F. Fox, IEEE Press, Chap. 7, 1999.

75. K. Bernstein, K. M. Carrig, C. M. Durham, P. R. Hansen, D. Hogenmiller, E. J. Nowak, and N. J. Rohrer, High Speed CMOS Design Styles, Kluwer Academic Publishers, Chap. 3, 1998.

76. Q. Zhu and M. Lingareddy "Low-Vt Devices Replacement for Domino Circuits," The 23rd International Conference on Microelectronics (MIEL-2002), pp. 597-601, Oct. 2002.

77. J. Pippin and G. Pham, "Willamette Domino," Intel Corporation, 2000.

78. P. Gronowski, "Issues in Dynamic Logic Design," Design of High-Performance Microprocessor Circuits, ed. by A. Chandrakasan, W. J. Bowhill and F. Fox, IEEE Press, Chap. 8, 1999.

79. "Nurlogic 0.18μm NLC_PLL266M Phase-Locked Loop," Nurlogic Inc., Aug. 2000.

80. "TSMC Macrocells Data Sheet PD22E1_4X – 66~166MHz PLL-Based Clock De-Skew Cell," TSMC, Nov. 1999.

81. "Clock, PLL & Clkgen Specifications," Chameleon Systems Inc., Nov. 2001.

82. "PLL Florplanning Guidelines," Nurlogic Inc., Aug. 2000.

83. S. Tam, S. Rusu, U. Nagarji, R. Kim, J. Zhang, and I. Young, "Clock Generation and Distribution for the first IA-64 Microprocessor," IEEE Journal of Solid State Circuits, Vol. 35, No. 11, pp. 1545 – 1552, Nov. 2000.

84. V. Gutnik, and A. P. Chandrakasan, "Active GHz Clock Network Using Distributed PLLs," IEEE Journal of Solid State Circuits, Vol. 35, No. 11, pp. 1553 – 1560, Nov. 2000.

85. G. Geannopoulos and X. Dai, "An Adaptive Digital De-Skewing Circuit for Clock Distribution Networks," Technical Papers of Solid-State Circuits Conference, pp. 400-401, 1998.

86. R. Senthinathan, S. Fischer, H. Rangchi, and H. Yazdanmehr, "A 650-MHz, IA-32 Microprocessor with Enhanced Data Streaming for Graphics and Video," IEEE Journal of Solid State Circuits, Vol. 34, No. 11, pp. 1454 – 1465, Nov. 1999.

87. Monolithic Phase-Locked Loops and Clock Recovery Circuits – Theory and Design, edited by B. Razavi, IEEE Press, 1996.

88. I. Young, J. K. Greason, and K. L. Wong, "A PLL Clock Generator with 5 to 110 MHz of Lock Range for Microprocessors," IEEE Journal of Solid-State Circuits, Vol. SC-27, pp. 1599-1607, Nov. 1992.

89. J. Alvarez, H. Sanchez, G. Gerosa, and R. Countryman, "A Wide-Bandwidth Low-Voltage PLL for PowerPC Microprocessor," IEEE Journal of Solid-State Circuits, Vol. SC-30, pp. 383-391, April 1995.

90. V. R. Von Kaenel, "A High-Speed Low-Power Clock Generator for a Microprocessor Application," IEEE Journal of Solid-State Circuits, Vol. 33, No. 11, pp. 1634-1639, Nov. 1998.

91. M. Sitkowski, "The Macro Modeling of Phase-Locked Loops for the SPICE Simulator," IEEE Circuits and Devices Magazine, pp. 11-15, March 1991.

92. S. Can and Y. E. Sahinkaya, "Modeling and Simulation of an Analog Charge-Pump Phase-Locked Loop," Simulation, Vol. 50, pp. 155-160, April 1988.

93. B. A. A. Antao, F. M. El-Turky, and R. H. Leonowish, "Mixed-Mode Simulation of Phase-Locked Loops," Proc. of Custom Integrated Circuits Conference (CICC), Paper 8.4, May 1993.

94. E. Liu and A. L. Sangiovanni-Vincentelli, "Behavioral Representations for VCO and Detectors in Phase-Lock Systems," Proc. of Custom Integrated Circuits Conference (CICC), Paper 12.3, May 1992.

95. A. Demir, E. Liu, A. L. Sangiovanni-Vincentelli, and I. Vassiliou, "Behavioral Simulation Techniques for Phase/Delay- Locked Systems," Proc. of Custom Integrated Circuits Conference (CICC), Paper 21.3, May 1994.

96. I. Young, M. F. Mar, and B. Bhushan, "A 0.35mm CMOS 3-880MHz PLL N/2 Clock Multiplier and Distribution Network with Low Jitter for Microprocessors," Digest of Technical Papers on IEEE International Solid-State Circuits Conference, pp. 330-331, 1997.

97. V. R. Von Kaenel, D. Aebischer, R. V. Dongen, and C. Piguet, "A 600MHz CMOS PLL Microprocessor Clock Generator with a 1.2GHz VCO," Digest of Technical Papers on IEEE International Solid-State Circuits Conference, , pp. 396-397, 1998.

98. D. W. Bailey and B. J. Benschneider, "Clocking Design and Analysis for a 600-MHz Alpha Microprocessor," IEEE Journal of Solid-State Circuits, Vol. 33, Nov. 11, , pp. 1627-1633, Nov. 1998.

99. H. Fair and D. Bailey, "Clocking Design and Analysis for a 600MHz Alpha Microprocessor," Digest of Technical Papers on IEEE International Solid-State Circuits Conference, pp. 398-399, 1998.

100. L. T. Pillage and R. A. Rohrer, "Asymptotic Waveform Evaluation for Timing Analysis," IEEE Transactions on Computer-Aided Design, Vol. 9, pp. 352-366, 1990.

101. P. J. Restle, T. G. McNamara, D. A. Webber, P. J. Camporese, K. F. Eng, K. A. Jenkins, D. H. Allen, M. J. Rohn, M. P. Quaranta, D. W.Boerstler, C. J. Alpert, C. A. Carter, R. N. Bailey, J. G. Petrovick, B. L. Krauter, and B. D. McCredie, "A Clock Distribution Network for Microprocessors," IEEE Journal of Solid-State Circuits, Vol. 36, Nov. 5, pp. 792-799, May 2001.

102. G. Northrop, et al. "A 600MHz G5 S/390 Microprocessor," Digest of Technical Papers on IEEE International Solid-State Circuits Conference, pp. 88-89, 1999.

103. A. G. Aipperspach, et al. "A 0.2mm 1.8V SOI 550MHz 64-b PowerPC Microprocessor with Copper Interconnects," IEEE Journal of Solid-State Circuits, Vol. 34, Nov. 11, pp. 1430-1435, Nov. 1999.

104. T. McPherson, et al, "760MHz G6 S/390 Microprocessor Exploiting Multiple Vt and Copper Interconnects," Digest of Technical Papers on IEEE International Solid-State Circuits Conference, pp. 96-97, 2000.

105. P. Hofstee, et al, "A 1-GHz Single-Issue 64-b PowerPC Processor," Digest of Technical Papers on IEEE International Solid-State Circuits Conference, pp. 92-93, 2000.

106. J. Silberman, et al, "A 1.0-GHz Single-Issue 64-Bit PowerPC Integer Processor," IEEE Journal of Solid-State Circuits, Vol. 33, Nov. 11, pp. 1600-1607, Nov. 1998.

107. "ClockStorm User Guide," Simplex Solution Inc., Feb. 2001.

108. I. Abu-Khater, "Basic Extraction Concept," Intel Corporation, March 1998.

109. N. D. Arora, K. V. Raol, R. Schumann, and L. M. Richardson, "Modeling and Extraction of Interconnect Capacitances for Multilayer VLSI Circuits," IEEE Transactions on Computer-Aided Design of Integrated Circuits and Systems, Vol. 15, No. 1, pp. Jan. 1996.

110. Q. Zhu, "RC Extraction in Deep-Submicron VLSI," Intel Corporation, Jan. 2000.

111. "AcadiaTM User Guide," Synopsys Inc., 2002.

112. "StarRC-XT User Guide," Avant! Inc., 2002.

113. "ElectronStorm Manual," Simplex Inc., 2001.

114. G. Wagner and J. Maiz, "Key Circuit Elements for ESD Protection on Process 650," Intel Technology Journal, pp. 4-12, Summer 1992.

115. G. Wagner, "Designing a Robust ESD Protection Circuit for Process 650," Intel Technology Journal, pp. 13-18, Summer 1992.

116. R. Aslett, "A Standard Cell ESD Protection Methodology," Intel Technology Journal, pp. 19-30, Summer 1992.

117. A. L. Fisher and H. T. Kung, "Synchronizing Large VLSI Processor Arrays," IEEE Transactions on Computers, Vol. C-34, No. 8, pp. 734-740, 1984.

118. D. E. Brueske and S. H. K. Embabi, "A Dynamic Clock Synchronization Technique for Large System," IEEE Transactions on Components, Packaging, and Manufacturing Technology, Part B: Advanced Packaging, Vol. 17, No. 3, pp. 350-361, 1994.

119. F. Anceau, "A Synchronous Approach for Clocking VLSI Systems," IEEE Journal of Solid-State Circuits, Vol. SC-17, pp. 51-56, 1982.

120. T. Gabara, W. Fischer, S. Knauer, R. Frye, K. Tai, and M. Lau, "An I/O CMOS Buffer Set for Silicon Multi-Chip Modules (MCM)," Proc. of IEEE Multi-Chip Module Conference, pp. 147-152, 1993.

121. "TSMC 0.18μm Logic 1P6M Salicide 1.5V/3.3V Design Rule (LV)," Taiwan Semiconductor Manufacturing Co., Nov. 2000.

122. Q. Zhu, "Chip and Package Co-design of Clock Network", Ph.D. Dissertation, University of California, Santa Cruz, June 1995.

123. W.C. Elmore, "The Transient Response of Damped Linear Networks with Particular Regard to Wide-Band Amplifiers," Journal of Applied Physics, Vol. 19, No. 1, pp. 55-63, Jan. 1948.

124. D. W. Marquardt, "An Algorithm for Least-Squares Estimation of Nonlinear Parameters", Journal of Social Industry, Applied Mathematics, Vol. 11, pp. 431-441, 1963.

125. Q. Zhu, "Method and Apparatus for Clocking a Sequential Logic Circuit", USA Patent No. 6037801, Granted to Intel Corporation, March 2000.

126. "User Guide of PathMill Static Timing Analysis Tool", Synospsy Inc., 2002.

127. R. Kumar, "Nosie Design and Analysis", Intel Corporation, 1997.

128. Q. Zhu, "Low Voltage Swing Circuits for Low Power Clock Distribution and Methods of Using the Same", USA Patent No. 5850157, Granted to Intel Corporation, Dec 1998.

129. T. Soyata and E. G. Friedman, "Retiming with Non-Zero Clock Skew, Variable Register and Interconnect Delay", Proc. of IEEE Intl. Conference on Computer Aided Design, pp. 234-241, 1994.

130. E. G. Friedman, "The Application of Localized Clock Distribution Design to Improving the Performance of Retimed Sequential Circuits", Proc. of IEEE Asia-Pacific Conference on Circuits and Systems, pp. 12-17, 1992.

131. M. Afghahi and C. Svensson, "A Unified Single-Phase Clocking System for VLSI Systems", IEEE Journal of Solid-State Circuits, Vol. 25, No. 1, pp. 225-232, 1990.

132. K. Bernstein, et. al., High Speed CMOS Design Styles, Kluwer Academic Publishers, 1998.

Index